Pocket
Surveying Buildings

Malcolm Hollis

Building Pathologist, Chartered Surveyor and
Professor of Building Surveying

Acknowledgements
The author and publishers wish to thank the following
for permission to reproduce copyright material:

HMSO; Housing Association Property Mutual (HAPM);
London Hazards Centre

Published by RICS Business Services Limited
a wholly owned subsidiary of
The Royal Institution of Chartered Surveyors
under the RICS Books imprint
Surveyor Court
Westwood Business Park
Coventry CV4 8JE
UK

ISBN 1842190741

Reprinted (with corrections) 2002, 2004

Typeset by Columns Design Ltd, Reading
Printed in Great Britain by Arkleprint Ltd, Northampton

Contents

Preface

Surveying buildings is an art, verifying the cause of the failure is a science. The surveyor's work involves a combination of both art and science.

No book can be a substitute for practical experience. This book is intended to provide a reference for the surveyor working within a building and is a companion to *Surveying Buildings*. When I wrote the first edition of *Surveying Buildings* in 1982 it was a third of its current size. In part this is because the subject has expanded, but it is also because of the effect of changes in methods and standards of construction as well as a legacy of errors within building construction.

I cannot carry a 500 page book with me when I inspect a building, which is why I have now written a pocket sized version. It should contain the minimum information needed during the inspection, whilst allowing reference to the illustrations and expanded text in the main book at a later time.

Building inspection is more than just looking at the visible surfaces of a building to see if there is anything wrong. It is also a process of verifying that the construction meets modern demands and that nothing of importance is missing. It may also involve a process of validating the quality of the performance of the building. That could include testing the building for the presence of unacceptable levels of dampness, testing the adequacy of the heating system, the electrical installation, the insulation or the drains. Whatever level of inspection has been agreed, what is important is that the work is done to the highest of standards and the report informs the reader of the implication of those limitations in the building that you have identified.

Malcolm Hollis,
London, March 2002

What is a Survey?

A survey is the inspection and investigation of the construction and services of a property in sufficient detail to enable a surveyor to advise what impact the condition and the circumstances of that property will have upon the client. The extent of the inspection must be sufficient to enable the surveyor to advise upon any future problems that may occur with the various components of the building. The surveyor must also be in a position to advise the client of where the property falls short of the requirements of modern legislation, and of any alterations that must be carried out to meet those requirements.

TYPES OF SURVEY

A survey may be required for a large number of reasons, and the investigation is usually a balance between the cost of finding out and the value of the information. The extent of the inspection may vary depending upon the type of work and the limitations of the instructions.

1. The sale of a property:
   (a) A report required by a prospective purchaser.
   (b) A report provided to the seller for disclosure.
2. Company sales or changes in ownership.
3. Valuation.
4. Maintenance management.
5. Pre-emptive maintenance.
6. Repair of failures.
7. Partial inspection and reporting – report upon part of the building.
8. Diagnostic report.
9. Condition surveys.
10. Dilapidations.
11. Access audits.

TAKING INSTRUCTIONS

Before any inspection there must be agreed with the client a statement of the actions and limitations of the inspection and of the content of the report. The client should be given time to consider a confirmation letter where the request is to report on part of the building or on a specific element of the services. Information sought prior to considering the offer of work should target the property, the limitations of the inspection, and the location of the building. The questions asked may include the following:

1. What does the client want to know?
   From this it will be determined whether the surveyor has to carry out:
   (a) a building survey;
   (b) an appraisal relating to proposed alterations of the premises;
   (c) an inspection so that the surveyor can advise if certain alterations work can be carried out;
   (d) an inspection so that the surveyor can advise if certain remedial work should be carried out and who should pay the bill;
   (e) an evaluation of the cost of rebuilding the property for insurance purposes; and
   (f) a limited inspection so that the surveyor can comment on a specific problem or the condition of a specific part of the building.
2. The date by which the information is required.
3. The approximate size, shape and age of the building that is to be inspected.
4. If the client intends to carry out any alterations if they proceed with the purchase? There is no point reporting on items that are to be altered or replaced.
5. Does the client require a detailed report on boundaries, outbuildings and the grounds of the property?
6. Does the client require an estimate of the rebuilding costs of the property? Is this needed for an insurance company or a mortgage company?
7. Details of lease, or other special information. If the property is leasehold then a copy of the lease should be provided.

The fee

Confirm the fee for the job. Also confirm a method of charging in the event that further work is commissioned, or the extent of the inspection is extended by a later agreement.

An instruction pro forma should be developed and completed by the surveyor or a member of the office staff trained to ask the relevant questions during a telephone enquiry.

Web sites

There is now a lot of information that is available over the web. It is useful to access information about the property before leaving to undertake the inspection.

Location: maps (and aerial photographs in some areas) are available from: www.multimap.com Type in the full post code and you can see where you are going.

Background: information about the site is available by entering the full post code at www.homecheck.co.uk The information is very general but may warn of the level of risk in relation to flooding or ground movement. Maps are included.

Flooding risk information is available from the Environment Agency. They move their web site around but access is usually available via www.environment-agency.gov.uk/subjects/flood/ Maps show the weak places and the risk of river and sea flooding. www.streetmap.co.uk (for location of the building or for street maps within UK based on street name, post code or a range of access points).

The Inspection

PUBLISHED GUIDANCE

The Royal Institution of Chartered Surveyors (RICS) has issued guidance to surveyors undertaking the building survey. The RICS recommends that the surveyor should question the building, those deficiencies suspected and seek the cause and extent of any problem that may arise.

HSV Report

Published by the RICS, it is now the dominant format for building reports. The HOMEBUYER Survey and Valuation Report's practice statements for England and Wales and Scotland are set out in the *RICS Appraisal and Valuation Manual* (the 'Red Book') in Practice Statement 11 and Practice Statement 11 (Scotland).

It has been suggested that the inspection for the HSV requires the same level of expertise as that for a structural survey inspection. This was quoted by Phillips J in *Cross* v *David Martin and Mortimer*. Within this first instance judgment no differentiation was made between the expectations of the inspection of the building survey and the HOUSE BUYER Survey and Valuation Report (HBRV). (The report format and name have been changed on a number of occasions between its introduction and the present date.)

The aim of the survey

* Identify the defects within the building and advise upon the consequences.
* Validate the performance in key areas, e.g. the adequacy or otherwise of damp courses, heating systems, etc.
* Verify the compliance of the building with such regulations that may apply, e.g. means of escape, Building Regulations.

Learning the trade

The surveyor should look. 'The first rule of surgery, gentlemen, eyes first and most, hands next and least, tongue not at all' (*Doctor in the House*, by Richard Gordon). The qualified surveyor should exercise the same degree of care, and commence the inspection with their eyes first and most.

* The survey inspection is of exposed and accessible parts of the building, unless the confirmation has agreed to a more intrusive investigation.
* Tests upon concealed elements may help to identify the presence of a problem but they are expensive.
* Tests should only be recommended if they will represent value-for-money.
* Any test should check or confirm what the surveyor anticipates to be a problem.

Horizontality and verticality
The surveyor must examine the horizontality and verticality of the components of the building.

* Check the horizontality and straightness of the ridge.
* Line up the walls by standing at one corner and looking along the line of the wall to view the far corner.
* Look along the line of a mortar joint to see if it is straight and horizontal for the width of the elevation.
* Walk the floors to see if they are level.
* Examine every junction, floor to wall, wall to wall, wall to ceiling.

Using your fingers

* Fingers are more likely to be able to pick up minor irregularities in a surface that is not lit by a strong cross light.
* Raking at mortar joints can often tell you if the cement-sand mix is too weak. The fingernail will be worn down by a sound or over strong mix, but will cut into a weak and too weak mix.
* Tapping surfaces can tell you about the construction and its stability by the sound made. Timber frame or dry lining sounds hollow, brick solid and blocks can echo, depending on size and weight. The varying types of construction should help in identifying the cause of any fractures and cracks which may be visible on the surface.
* Tapping metal or clay pipes with a coin will produce a sound that may help diagnose the condition: a dull thud if blocked, a jangle if cracked as opposed to the clear ring of a sound pipe.

* Tapping wall plaster or tiles can indicate if there has been a loss of adhesion – the hollows to the back of the plaster or render echo and give a rattle sound.

The tapping of timbers may suggest the condition.

* Where there are indications of fungal decay a different sound is produced to that from undamaged timber. Because much of the rot or decay may progress through the middle of the wood of a beam, the external surface may be free of any sign of failure.
* In defective timber the sound may be slightly hollow, have a dull thud or a dead sound, as opposed to the unaffected wood, which gives you a substantial note and sounds solid. The sound material has a ring in the note produced by the vibration of the wood when struck. The diseased timber, which has damaged fibres and cells, is unable to produce the same note and sounds dull.

Feet and floors
Note any creaking and movement of timber floors. This could relate to structural weaknesses but is always a cause of complaint.

* Gently bouncing on your feet will reveal the springiness of the floor joists. If you find movement, point out what it could mean. The result may suggest component failure, deterioration due to damp. Undersizing of joists or risk to embedded plumbing.
* The movement may be the result of overloading, expansion in the components of the building, dampness or heat?
* Diagonal cracking in a wall is unlikely to be thermal – a vertical split down a wall may be. Parapet dislocation – particularly where it is stepped – may be. Horizontal displacement is unlikely so to be.
* Identify any variation in level by walking around the floor surface. Check with a spirit level.

Using your nose
Various material failures have very distinctive smells. The nose's ability to identify a smell is only possible on a comparative basis. The various smells that can indicate dampness within a property are easily recognized with experience.

* Dry rot has a musty, spicy tinge, a dull smell that includes just a hint of dirty socks.
* Dampness resulting from hot water pipe failures smells like burnt water, being slightly humid.
* Dampness smells of the material that the moisture has caused to decompose. It is often a smell of wet newspaper.

Dampness
The moisture meter can give you information that, if interpreted, may suggest a risk of dampness. It is for the surveyor to confirm that the information does mean dampness and then to diagnose what caused the moisture.

The material of a wall has a limited ability to allow water to climb. Rising damp will rise further in plaster, particularly if it is carlite, and will appear at higher levels if there is a salt content in the wall or plaster. Pressure may push rising water higher where a solid concrete floor has been added or damp retardant wall finishes have been applied.

Reporting on the consequences
Eliminating the cause of water penetration is essential. Dry rot is an almost inevitable consequence of ongoing water penetration if timbers in poorly ventilated locations become damp. Our job is to warn what might happen and then to recommend what steps the client should take to identify the extent of the infestation, and when those steps should be undertaken. Wording such as:

> Before you commit yourself to this purchase it is essential that the extent of this outbreak be identified and the cost of its elimination and subsequent repair calculated. This will involve intrusive examination. Without this knowledge the value of the building cannot be assessed.

Warn, recommend and then advise. Without exposing the rot no one can report accurately upon the extent of an infestation.

New technology
Video equipment and cameras (both film and digital) are a useful method of recording evidence available at the time of the survey. The digital camera has improved the surveyor's ability to see around corners. By pushing the camera into gaps, up chimneys or outside windows, pictures can be taken of inaccessible locations. The rear view panel on the camera enables a quick check to be made during the inspection whilst the detailed inspection can be made on returning to the office. The image will help the reader to understand the report.

There is insufficient time during a routine survey to check a property with all the equipment that is available. But if the intention of the inspection is to determine the cause of the damage then the extent of testing will increase, as will both the time required and the cost of the work. If the surveyor has been asked to diagnose the cause of a failure, and to advise upon the cost of correcting the problem they should be aware of the equipment available and the information that it can provide. For example:

* Infra-red thermography can locate deficiencies in cavity wall insulation.
* Impulse radar can locate fissures within walls 3m thick.
* Sonar enables a profile of solid material to be created so that voids can be identified.
* Fibre optics enable images from concealed parts of the construction to be examined.
* Sibert microprobe measures the resistance of timber drilled, from which quality and age can be determined.
* Water Level to determine levels around and within a building.

Length of survey
There is no set time for a survey. It will depend upon the building, the conditions and what is discovered. What is essential is that sufficient time is allocated and used in the inspection and report drafting. The quality of an inspection will be affected if there are either interruptions or disruptions.

The inspection should take enough time for the surveyor to focus upon the building. Looking at a building is a complex thing, one is looking at the condition of what is there and one is looking for the defects that are common in that type of construction. One must concentrate upon what could be seen.

Condition of the building
The condition of the building is relative. It will affect the client, as prospective owner, leaseholder, or occupant of a building. The effect on the homeowner of essential repairs will vary, depending upon, for example, their health, the quality of the contents that could be damaged or their sleep if they work nights. A weak roof may fail sometime in the future. For the average person that may mean waiting for it to happen and dealing with the consequences at that time. For the special occupant it may be essential that the roof does not fail. Some knowledge of the client is important so that special circumstances can be taken into account

Orientation of the building
For certain religions, compass directions are important. For most purchasers knowing where the sun will shine is

relevant. It also helps to define the parts of the building being discussed in the report.

Leasehold properties
The survey of leasehold property, be it a flat or business premises, requires a different approach. This time it is not just the condition that is to be recorded, it is the liability for the remedy that needs consideration. The failure of the external walling, where water penetrates into a flat, would be noticed during a survey. The repair liability could be that of either the landlord or the tenant, or be unspecified. The obligation to pay for the repair may lie with the tenant, through a service charge, if the landlord does the work. The defect may not be a repair and therefore, although the tenant has water pouring into their accommodation, the landlord has no obligation to remedy the situation.

Looking beyond what is there
To look for what should be there, the surveyor needs to be an expert in construction. All building forms have their own requirements; they may be specific to a concrete frame, to the cladding or glazing of a frame construction, to a warm or cold flat roof or any other form of construction. That means that the type of construction must be identified so that the surveyor is able to determine what should be there and whether anything is missing. The condition can only be interpreted when related to the method of construction and the materials that have been used.

What should not be there?
The inspection will seek out what should be there but is missing. What should not be there may include electric switches in a bathroom. The 220/240 volt system in the UK and much of Europe differs considerably from the 110 volt system in the USA. One can kill and the other is unlikely so to do. The demands upon an electrical installation will vary. But the rules are set down for each location and must be followed because people's lives may be at risk.

Checklists
A checklist does help to reduce omissions from the inspection. The checklist set out below aims to provide a framework for the inspection of a simple residential property.

Triggers
Looking does benefit from triggers. One trigger could be 'differences'. One looks at the building to record the differences between it and any adjoining building. Another label could be 'movement'. Knowing that each

component of a building will move, and that the individual movement of different materials will also differ, one looks at the building for the junctions between components to see how provision has (or has not) been made for such dynamic change.

The most common label will be that of 'comparison' – the comparison of this building with the appearance and complement of similar well-built examples. The image of one is carried to the actuality of the other, and differences are noted.

Looking for evidence
Looking is not just a function of seeing defects in a building, it includes searching for the links between defects and components that may explain the full extent of a problem. Identifying defects relies upon familiarity with risk locations built up from previous inspections. The method of discovering deficiencies during past inspections will be repeated in future work. That experience suggests what to look for and what to look at. This can be a risky process, as unfamiliar failures are less likely to be identified.

Questioning the vendor
It is natural for the homeowner to be defensive when they are asked about the problems they may have had with the building they are selling. The new Seller's Survey, which is unlikely to be introduced before 2005, will lead to the surveyor interviewing the vendor to gather information upon the quality of the building. Think about the method of asking for information. Ask about:

• The length of time that they have been in the building and whether they are moving far. That would establish the level of knowledge they may have, and whether the reason for moving could be because of the cost of maintaining the building.
• The level of gas, oil and electricity bills.
• Recent maintenance work – who they use and how regularly. 'When did you last decorate?' – an innocent question for which there should be no need to conceal the true answer.
• Make statements for them to respond to, such as 'I see that you have had woodworm problems – there is a tin of treatment liquid in the garage'. The vendor is forced to acknowledge the accuracy of the observation.
• Whether the building has required much maintenance.

The process must allow you to achieve reliable information without letting the vendor know that they are being closely questioned, or placing them in a position where they have to admit to a lie.

Accuracy of diagnosis
Consider what level of accuracy can be achieved whilst carrying out an inspection. You cannot state categorically that all defects in a building have been located. If one accepts that the survey is limited, one needs then to consider by how much it has been restricted. In order to get closer to the extent of restriction, it is useful to consider the problem by working in reverse. The purpose of the survey is to try to find out the presence, or the risk of the presence of particular defects within the building. If one starts with the concept that there is a risk of, for instance, dry rot in the property, how do you establish that there is no dry rot in the building? If you only examine 7% of the building and want to discover the risk of dry rot occurring, (in areas where there is timber, dampness, and limited ventilation) one can see that it is impossible to certify that there is no dry rot within the property. Short of carrying out major surgery to determine the condition of all of the building, only the absence of timber, damp or limited ventilation in any part will enable confirmation to be provided that no rot exists.

In order to obtain greater accuracy, damage and substantial costs must be incurred. The accuracy of the surveyor's diagnosis, which relies upon experience to evaluate the risk of failure, will be increased by exposure to errors in both construction and defect analysis. The level of accuracy in the inspection will be influenced by the amount paid, which affects the extent of that inspection and the time available. It will also be influenced by the amount of the property that is available for inspection. The surveyor's knowledge and experience will determine the level of accuracy within the parameters identified by time and money.

Without a detailed inspection, the surveyor's judgment can only be general and can only stress the possibility or probability of a failure based on their knowledge and experience of similar properties. It would be arrogant for a surveyor to indicate that they have determined the full extent of all the defects within a building.

Looking

How do you read an image? Do you, as with a book, start at the top right (or left) and run your eye down the building? We receive little training in the task of looking.

Seeing

There is a difference between seeing and looking. Seeing is the result of the mind interpreting the observed element or composite. To see one must first understand. One may look at an abstract painting but not see it. To see one must first look, and then recognize. Recognition is the gateway to seeing, because one cannot see without understanding – otherwise it would be just looking.

Our recollection of images enables us to place labels on what we look at. But to understand, reliably, the image, we must learn to look, to observe, to challenge before we accept the meaning of what we see. To see before we look is judgmental, leaving us open to confusion, open to being deceived by what we think we see.

Experience shows us where one should make sure that one has looked. Knowledge may develop the risk analysis of the observed, and will direct one to areas that require close examination. For example, inadequately sealed flashing directs one to examine the areas alongside and below for water entry.

The forces that propel water from the outside into the interior of the building include:

* Capillary forces – the characteristic of water to flow through masonry or hairline cracks, usually by thermal catalysts.
* Kinetic forces – the kinetic energy of a wind-driven rain will force water into the depth of the wall.
* Pressure differential – net pressure differentials caused by ventilation and air conditioning systems may cause water to be sucked into voids created by the pressure differential.
* Gravity – under gravity, water can drip in through imperfections in flashings, gutters, roofs and parapet walls.
* Surface tension – water will tend to follow a wet surface. It will flow around corners and edges such as in soffits, shelf angles, and loose laid metal flashing. Add a touch of wind and it will all happen that much faster.

WHAT SHOULD THE SURVEYOR KNOW?

* The composition of the soil. See soils maps produced by the British Geological Survey.
* Construction weaknesses – explained in information sheets published by the Building Research Establishment (BRE).
* British Standards or Codes of Practice. The fourth grouping relates to the rules of statutory bodies.

- The RICS HOMEBUYER Survey and Valuation Service Practice Notes PSA 12 and PSA 12 (Scotland), *RICS Appraisal and Valuation Manual* (the 'Red Book') (1997 and 1999). Coventry: RICS Books.
- Guidance on radon, mining, Mundic, weak or limited forms of non-traditional construction.
- The Building Research Establishment Guidance and Advice.
- British Standards, e.g. BS 5250 dealing with roof ventilation equivalent to a 10mm slot on two opposite sides of the roof where the pitch is in excess of 15°C, but a 25mm standard for roofs of a lower pitch.
- Statutory Bodies Regulations, e.g. Gas Board regulations relating to flue outlets and required clearances. Balance flues at least 300mm from any obstruction, opening or projection except from surface facing terminal or another flue, or 900mm from a plastic gutter. (See also Building Regulations F27, and British Standard 5440).
- Institute of Electrical Engineers Regulations, 16th Amendment, specifically for requirement of earth bonding and relationship between electrical fittings and plumbing and insulation material.

Carrying out the inspection

Before starting the inspection spend sufficient time getting to know the building. That involves going inside and out, to verify, check or note aspects that challenge each other. The inspection is the collection of information. The notes are items from which the inspection will be prompted and the report drafted. The report is the considered opinion.

CHECKLIST IN PREPARING FOR A SURVEY

A checklist can direct the surveyor to answer only the questions listed, as opposed to the further questions that may be posed by the building.

PREPARATION

- Confirmation of instructions to be sent before inspection.
- Note any agreed specialist testing.
- What is the size of the property?

- What type of property – is it an office or a shop?
- Are there any fire certificates, or other consents that are relevant, that you will need to see?
- Is the building listed or in a conservation area?
- You must have sufficient knowledge for this inspection.
- How soon does the client need the report?
- You should have some knowledge of the requirements of the client. If not, assumptions should be stated.
- Advise the vendor of the probable length of the inspection.
- What does the client propose for the building?
- Has sufficient time been allocated for the survey?
- What is the full address? Can you find it?
- Are there any guarantees?
- Has the client seen the property?
- Do they know if it is suitable for their purposes?
- Do the vendors know how long you will be there?
- Do you have to arrange any specialist tests?
- Have you sent a confirmation of instructions letter?
- When are you going to do the job?

Leasehold properties

- Obtain certified copy of the lease.
- Check colour on plan of demise in lease.
- Obtain copy of management accounts.
- Is there a caretaker's flat to inspect?
- Can you gain access to the rear, roof, etc?
- Are there any special dates for redecoration, or when major repairs have to be carried out by, referred to within the lease?
- Is the property on an estate where the properties must always be maintained to a specific high standard of repair?

The visit

- Ensure that any equipment to be used is in good order.
- Has the damp meter been calibrated?
- Take a list of any special instructions from the client to the property.

- Do you have adequate identification for the benefit of the occupant?
- Confirm access arrangements and availability of the key if it is to be collected.

The adjoining properties

- Are there any failures in similar properties in the neighbourhood?
- Are there any major works being carried out to a similar property in the area?
- What changes have been made to similar properties in the area?
- What is the condition of adjoining properties, particularly if they are attached?

The soil

- What type of ground is the property built upon?
- Is there any history of settlement or contamination in the area?
- Are there any trees close to the building, or its drainage, that may affect the property?
- How wet is the ground? Are there any watercourses in the area? Is there any history of flooding? How well is the area drained?
- What type of foundations would you expect the building to have? Are they likely to be deep enough in clay or adequate for the ground under the building influenced by the circumstances found?

The grounds

- Can you identify or locate the boundaries?
- What is the condition of any fence, wall, hedge or other boundary demarcation? Are there any special reasons for boundaries to be in very good condition (e.g. for animals to be kept on the premises)?
- Are there any outhouses or sheds? Have you inspected them? If not, say that you haven't.

Outside influences

- Noise – of aircraft, neighbours, particularly in flats.
- Noise from local buildings at the time of your inspection (such as factories, schools, or traffic).

* Smells from factories, traffic, farming or the ground.

THE INSPECTION

Exterior

* Record the weather at the time of your inspection and any limitations to what you can see of the exterior.

The roof

* Record the type and appearance. Is the client likely to have seen it? Do you need to include a sketch or photograph to explain the roof layout?
* Record the material of the roof covering and its appearance. Note any untrue or uneven surface.
* The roof covering. Is the slope angle correct for the material used? Is the surface the original material? If it has been changed, has someone thought through all the construction changes that may have been needed, e.g. strengthening of timbers?
* Are there adequate cover flashings where there is an upstand to a flat roof?
* How has the covering lasted on other properties? Are there any disadvantages to the material that has been used?
* How is it fixed? If you can't see, could the fixing be deficient, e.g. nail sickness, peg tiles, peg failure, failed tile nibs, poor laps?
* What is the difficulty of repair if a remedy is required? Can scaffolding be erected from within the boundaries of the premises?
* If condensation occurs below the roof surface, what will happen? Is the roof adequately ventilated?
* Are there any openings through the roof surface? How have they been sealed? Can water penetrate?
* What is the condition of any flat roof surface? Is it a cold or a warm construction?
* What happens in driving rain? What happens in snow?

Chimneys

* Are there any defective flashings? What is

the material of the flashing? What is the likely life span, and what condition is it in?
* What is the condition and security of the pots? Are they special, do they have to be replaced in the same design?
* Are the flues open, or have they been ventilated? If not, are there pots on top, or have they been capped?
* How have the chimney openings been dealt with within the building? Are the chimney breasts continuous within the building? If not, how have they been supported?
* What are the special needs of the flues, the boiler, the wood fire, the coal fire, ventilation, etc.?
* What regulations may apply to the location of the flue, and the fuel used?
* Are there any damp stains close to the outside of the chimney?

Rainwater disposal

* Are there any signs of leakage or failure?
* Can you see all the surfaces to be able to record their condition?
* Is the capacity of the system adequate? Is it effective? Is there a felt dressed into the gutter?
* Is the gutter placed clear and beyond the face of the wall below?
* Is the gutter clear of debris?
* Does the roof felt discharge into the gutter or does water run behind the face?
* What materials have been used? Are the materials compatible? Are they prone to corrosion?
* Is there a history of failure with the disposal system that has been used, or with the materials used?
* Are the materials still available in the event that repairs are needed?

Eaves detail

* If water penetrates into the wall will it do any damage? Are there any timbers bedded into the wall?
* Eaves. Will water run off into a gutter, or behind it? Are there any stains on the wall to give you a clue?

Walls

* Construction: solid or cavity? What are the components of the wall and the implications of the construction?
* Are there adequate movement joints and what is their condition?
* Has the wall been rendered? If so, why?
* What is the condition of any render and does it need repairs? If so, what damage may already have taken place? How does one inspect to find out?
* Are there any signs of cracks? If so, how serious? How big are the cracks? How big are they if you add all the cracks together? What is the cause – thermal, drying out, component failure settlement, and subsidence?
* Are the cracks in a particular direction? Does direction and frequency suggest cavity tie failure, chemical attack, or movement in the foundations or beams or supports above the ground?
* Are the walls straight or are there any bulges? If so, how far has the wall bulged? How much disturbance is there to the internal door frames, window jambs or door jambs?
* Where are the window frames located in the wall? Does that suggest timber frame construction? Is the junction with the sill adequate?

Windows

* What is their condition?
* What is the condition of the glazing, double-glazing and the sealant used? Inspect the putty, the sill details, and the materials.
* Is water likely to bypass the glass or the frame? If so, where will water get trapped? If it is trapped, how much damage can it do, and has it done?
* How weather-tight is the design? How good is the seal to the head or side of the windows?
* What condition is any sub sill? Does it project beyond the wall face; are all drips clear of blockage? Is there any damp penetration?
* Do the windows open? How secure are they against forced entry?

Damp proof course (DPC)

* Can you locate it? If so, is it 150mm above ground level; is it bridged; where are the weep holes around the dpc? If below dpc it may be timber frame construction.
* If you assume presence of chemical injection dpc, will it work, has it worked, and have you checked? What is the moisture level of the brick joints above and below the line of the dpc?
* Are there any airbricks? Should there be any? Are they clear? Is there adequate ventilation?
* Will water, vermin or other undesirables enter through any opening to the outside of the building?

INTERIOR

Loft space

* Can you get into it with your 3-m ladder, or is a drop down ladder supplied? If not, advise that no inspection was possible and recommend that access be made. Should that be inspected before they purchase the building?
* How has the loft access been formed? Has it damaged the strength of the roof, is there a security problem if the roof void is not self-contained?
* Is the roof void fully enclosed, or are there openings in any party wall between this and another building? If so, can they be closed, for security or as a fire risk?
* Is the roof close boarded? If so, can any inspection of the underside of the roof timbers or roof covering be made? If not, advise upon the limited access.
* Is the roof felted beneath the roof tiles or slates? If so, advise that no inspection of the battens and roof covering can be made. Tap the felt to hear if any tile nibs have failed. Is there an adequate hollow for water to run past the tile battens? What is the condition of the felt, is it laid with the lap directing water outwards? Is there any damp at eaves level?
* What is the condition of the roof rafters and ceiling timbers?
* Is the roof frame adequate? Is there diagonal

bracing if it is required? Has the frame been strengthened if a heavier roof covering has been provided? What is the purlin span, and is the timber size adequate? (Maximum purlin span is about 2.5m.)

* If they are modern timber trusses, what is the condition of the metal nail plates? Has the galvanizing failed?
* Are the trusses in line and vertical?
* Are there lateral restraint straps, and if so are they secured to at least two trusses, secured to the gable or party wall face, and any gap between truss and wall solid blocked?
* Are the trusses at regular spacings, usually 600mm?
* Are there water tanks, and if not is there a condensing boiler? What is the support given to water tanks, and does that support span three timber trusses? Do the tanks have covers, does the ball valve work and is the water clean? Is the overflow pipe supported, running downhill and are the pipe joints sealed? Are the water pipes and the tanks adequately insulated?
* Is the roof ventilated? Has the ventilation been blocked with insulation? Is there adequate insulation in the roof? Was there any condensation on the underside of the felt, and if so, has it damaged any timbers or ceiling plaster?
* Is the roof insulated? Is thickness less than 200mm? Is it adequate? Does the insulation block the free ventilation of the roof void? What action should be taken?
* Is there any wiring in the insulation within the roof? What is the condition of the wiring? What type of wire is used?
* Have there been any animals or other wildlife in the roof void, such as bats, bees, wasps, pigeons, rats or mice, or squirrels? What damage have they caused, or could they cause? Can they be eradicated?

Interior of the building

* Damp checks – have you tested under windows, all walls, to exterior to internal partitions to ground floor, to skirtings, to the top floor at the back of gutters, internally where the external roof joins onto external walls?

- Is the plaster on plasterboard or timber laths? If on laths, point out the expected durability and the weight of plaster that will fall when it fails.
- What is the quality of the surfaces – is the plaster sound, is it loose, has it pulled from backing, will it survive redecoration? If not, advise upon the extra cost that will be incurred when internal redecoration is carried out.
- If the walls are dry lined, inspect the party wall carefully to determine whether you consider there to be a risk of excessive noise transference between buildings.
- Are there any cracks? What has caused them? Are they serious? Do they align with any external cracking? If so, there is a probability of ground movement.
- Wooden surfaces – are they free from beetle or fungal decay? Can you be sure? Are the symptoms there?
- Is there condensation? If so why, and what do you do to eliminate it?

Doors

- Do the doors open and close? If they bind, what may have caused this – damp, movement, seasonal variations, no central heating?
- Have the tops of the doors been altered to make them fit a distorted frame?
- Consider security implications for external doors.

Floors

- How are they constructed? Timber suspended, solid concrete or suspended concrete beam and fill?
- Timber floors: are the support joists bearing on a damp wall, or could the wall have been damp in the past? If so, what is the risk of concealed timbers being defective? Is there ventilation below the lowest floor level?
- Timber floors: are they supported on the party wall? If so, what is the risk of sound transference through the party wall? Listen during the survey for any sound transference to suggest a problem.

* Particleboard flooring on timber joists: these are usually tongue and grooved. Are the short edges supported on joists or noggings? Can you verify by presence of nails or by use of a stud-detector the presence of timbers beneath? Are the nail heads visible 3mm wide? Is there an indication that the tongue and grooved joints have been glued? Is there 10mm clearance around the edges of the room below skirtings?
* Concrete beam and fill: how is the floor placed on the supports? Is it particleboard on insulation? If so, are there any creaks as you move about on the floor? Test, and if present, warn of their presence and the high cost of eliminating the creaks – currently about £4,000 for a small house. This is not a defect covered by the NHBC/Zurich warranties on new houses. Is there ventilation below the floor supports? The reinforcement is at risk if ventilation is not provided.
* Solid floors: examine the gaps below skirtings all round the ground floor to see if there is any risk of movement in the floor slab. Research to see if the building inspector is aware of slab failure. Check the surface level to see if there is any lifting towards the centre that may suggest sulphates expanding the slab or the hardcore below. Tap the floor to see if there are any hollow areas that may suggest that the slab is dropping or that voids have been created by the hardcore being washed away.
* Solid floors: can you trace the damp course? Do you know if a membrane has been placed below the screed or the concrete slab and the screed? Do the damp checks you carry out support the idea that the damp course or membrane is working throughout the floor?
* Solid floor with cement screed: test the floor screed for level of dampness. If dampness is above the level anticipated, is there a risk of a plumbing failure allowing water to seep into the screed?
* Services in a floor screed: the water pipes must be sleeved or be tape wrapped – are they? Are the copper pipe joints encased by the screed? (They should not be.) Is there adequate cover over the pipes of at least 25mm?

- What is the quality of the screed? Can you rub it with your fingers, is it soft and friable? This may suggest a poor mix. Can you see cracks in the surface, particularly close to door openings? This may suggest movement in the building or in the floor slab independent of the building structure?
- Are the floor surfaces level? If not, where and to what extent? What could be the cause, the extent of the damage and the future risk for the building?
- How much damage do you think the problem could cause?
- Does any surface problem, even though concealed by carpet, suggest that further examination is required to determine the full extent of the existing problem?

SERVICES

Water

- Copper pipes should be clipped at a maximum of 1.8m both vertically and horizontally. Pipe bends should have no crinkled surface on the inner side of the bend. No pipes should be in contact with cementatious material.
- Insulation should be provided to all water-carrying pipework that is in areas that are not heated.
- Overflow pipes must fall down towards point of discharge in a conspicuous location, and be supported at not more than 900mm centres.

Electricity

- Electric cables must not be in contact with copper pipes except for special earth bonding. Wiring must not be in contact with polystyrene insulation.
- Electrical sockets in new build should not be less than 150mm above the floor or a kitchen worktop, or be within reach of a basin or sink, or within 2.5m of a bath or shower.
- The main water pipes, gas pipes and central heating pipes must be earth-bonded with copper conductors which are permanently labelled. The earth cable is yellow with a

green strip and the labels are usually metal fixed to the connector.

Heating

- Balanced flue boiler terminal should be within 300mm of opening above or 600mm of opening below. The flue should not be within 750mm of painted eaves unless metal protection has been fitted. A guard is required to cover the external outlet where it is within 2m of the ground.
- Where there is a conventional flue, the outlet must not be within 600mm of a dormer or parapet wall or less than 1m above the eaves, except where the roof pitch is below 45°C.

Reporting

COMMUNICATING

The report must advise the client about the implications of the faults that were found. Because the report is your considered opinion, there will be times when the edges between inspection, diagnosis and reporting may become blurred.

You must consider the defects found, the risk of problems occurring in the future together with those matters that may be a defect in the mind of the client. The report must both identify the defects found and advise upon the risk of defects being present.

Before the report is written it helps if the surveyor can sit back and reflect upon the building. If it were going to be bought by your best friend, your nephew, niece or your own child, what would you say to them?

Defining defects

The definition of a defect, as given by the Norwegian Building Research Institute, is 'unexpected expenditure incurred by the client following taking possession of a property'.

The defect is determined by the client and not by a building professional, and it is identified by unforeseen expenditure.

Maintenance costs

The average expenditure on maintenance of a home is over £1,500 per year. Leasehold commercial or residential property is more expensive to maintain. The value of flats is affected by the terms of the lease. A mortgage valuer is able to make certain assumptions as to the lease, even where no knowledge has been imparted, but must examine and advise upon the maintenance of leasehold-managed properties. The valuer's report will include the estimated current level of service charge. Poor maintenance adversely affects the value of the building. The remedy required to eliminate defects may be outside the landlord's leasehold liability.

Warning the client

We either report upon the defect that we know exists, or express a clear warning where we believe that a defect may develop. That warning must contain several elements:

* A simple statement of the perceived problem.
* A warning of what might happen.
* Your opinion: what do you really believe is the case?
* Your advice: what do you think your client or customer should do?

Summary page

The report will contain clear statements where you believe that there may be a problem. I suggest that the heading for that summary should include the following items.

* The building type – history of defects in this type of property.
* The work required prior to the client committing themselves to the purchase.
* The work required prior to occupation of the property.
* Work that should be carried out in the immediate future.
* Advice:
  ◦ How bad could it get.
  ◦ What work is required to remedy the defect.
  ◦ How bad do you think it is.

Problem solving is a process that requires the investigator to come up with ideas to determine the cause of the symptom discovered.

SEQUENCE OF PROBLEM SOLVING

* Define problem.
* Evaluate ideas.
* Brainstorm.
* Identify range of options.
* Test options and dispose of those which are not appropriate.
* Remainder are the possible solutions.

- Do not opt for a single cause.
- Test your hypothesis to prove it wrong.
- If you cannot disprove your idea of the cause, it may be the cause.

This sequence does not establish that this is the cause because there may be more than one fault that is creating the problem.

HOW TO REPORT SAFELY – A CHECKLIST

CUSTOMER EXPECTATION

Confirm the instructions; recheck that you have done what you undertook to do, and explain why you may have deviated. What did your client expect? To know what the condition of the floor under the carpet was? What could you see by easing back the edges? You can make an educated guess based upon this information, but not a statement of the absence of a defect.

TESTS FOR THE REPORT

- Should you have seen the fault?
- Were you negligent in failing to report upon the fault?
- Could you have recognized that the circumstances might have deteriorated, and if so, was an adequate warning provided?
- Did you fail to warn that the fault, or the embryonic stages of the fault, could develop, and was there an adequate warning or advice to inspect further?

INSPECTION

- You must have identified the construction and taken into account all knowledge available to the profession about the limitations of any of the components or the form of construction.
- You must have made a thorough investigation and have recorded the results of all tests and checks undertaken in field notes made at the time of the inspection.

DEFECTS

- Evaluating the significance of a defect is to a degree an essay in speculation.
- The courts will rely on published information available at the date of the report.

ADVICE

A surveyor must consider and advise the client upon:

- the likelihood that the damage will get worse;
- the extent of work required to eliminate the defect;
- the risk that more serious defects may occur; and
- ways of limiting the risks, e.g. further investigation, etc.

ALTERATIONS

If alterations have been carried out to the property, the surveyor must consider and advise the client on issues such as:

- Could the original design have been damaged by the alteration?
- If the original building design was not suitable for alteration, could you have ensured that the building would perform efficiently in the future? Was the work done properly?
- What should be done to check that any alteration has been correctly carried out and will be secure for the future?
- If you are unable to confirm the adequacy of the alteration, you should express a warning and recommend further research.
- If you are not sure of the adequacy of the work that has been carried out within the building, express a clear warning and recommend further investigations.

HSV

- The inspection for the HOMEBUYER Report and Valuation requires the same level of expertise as that for a structural survey inspection.

REPORT

Absence of evidence is not evidence of absence. Just because the surveyor did not see any dry rot, beetle, cracking, damp, chemical attack, asbestos, contamination, radon, or Mundic does not mean that they are not present.

- Do not conclude what you cannot substantiate.
- Set out a clear warning of the risk of more serious defects developing, as well as a clear warning of the seriousness of those defects that have been found.
- Recommendation should be set in the context of time, i.e. 'before exchange of contracts'.
- Indication of cost helps to clarify the serious nature of any fault that may have been found.
- The limited nature of the survey, being an inspection of the available surfaces of the building, results in the majority of the advice being limited to the use of either 'probable' or 'possible' when referring to defects being present within the building.
- Professional advice is less the achievement of the correct answer but rather the analysis, evaluation and selection of the appropriate option or options.

Specialists

Where there are aspects of the building outside the inspector's knowledge base, then the detailed advice should be referred to a specialist (for example, electrical test or asbestos analysis).

Exclusion clauses

There is a difference between a caveat, a limitation, and an exclusion clause. An exclusion clause should be declared within the contract that sets out what will and will not be done. Some limitations to the inspection may be suggested within the contract or imposed by limitations discovered during the inspection. A caveat is a warning of what may be not undertaken within the inspection. It is quite correct for the surveyor to advise where certain parts of the building could not be inspected. That does not mean that the possibility of a risk of failure in the concealed area will not be considered.

In the past ten years the main changes made in the survey of buildings have been in the styling of the reports by the use of standard words. This may protect the surveyor but does not help the customer to understand the report. There is a missing link in this type of reporting and that is the failure to see the connection, for example, between comments on corrosion and water, dampness and decay, or location and weather patterns.

The use of limitation clauses is an essential part of the report. The surveyor will have confirmed the instructions, but it is only after having completed the inspection that the extent of the work and any limitations can be confirmed. By contrast, the use of caveats in the report often is an indication that the surveyor has undertaken less than had been contracted.

Costings

References to the cost of repairs in the report should be qualified as being a rough estimate. 'The prices included within this report are for guidance. You are advised to obtain competitive quotations from local contractors before committing yourself to the purchase of this property.' BCIS have introduced guidance upon repair costs on a subscription basis.

Our reports should indicate the level of repair cost. Is there a simple solution to the defect in the building? What would happen if you did nothing? If the building has achieved an equilibrium there may be no reason to spend the money on a repair. If this is suggested, do make sure that you point out that this may affect a future sale because the defect remains. This influence on the value of a building will enable us to evaluate the effect of the defect. Whilst the cost of repair is not the method of assessing the variation in the value of the property with the defect, the cost of repair is a function in deciding by how much the value will change.

We can be creative by identifying the minimum repair cost. We are a thinking profession. Provide a balanced opinion of the options available and recommend the best solution, both cost-wise and work-wise. That inventiveness could result in our indicating how to recover the cost of repairs through grants, tax relief, or insurance.

Conditions

The report should:

- Include a copy of the agreed instructions.
- Follow a regular pattern so that the client may retrace the steps of the surveyor and identify the failures.
- Clearly identify the property.
- Set out the weather conditions prior to, and at the time of, the inspection.

Content

The RICS has published guidance notes. They have suggested that the surveyor should, in reporting upon the survey:

- state the discovered facts;
- describe the elements which could not be identified and indicate why;
- include recommendations for further enquiry or investigation prior to commitment to purchase;
- describe defects and disadvantages in relation to contemporary standards and to those applicable to the period of construction, and the likely consequences of non repair; and
- make recommendations in respect of the timescale for necessary work and advise on any long-term implications.

Conclusions

The report should clearly set out the conclusions that the surveyor has reached relating to the property.

Pro formas

Many surveyors inspect similar properties that are virtually identical and find a standard approach very helpful. There is a considerable risk in the use of a standard framework because the surveyor may feel the inspection is about completing a form. They may fail to investigate the cause of each problem.

Reporting upon the components in isolation may result in a failure to report upon the links between each element, and the damage those links or connections cause. A survey report must be the reflection of the consequences of interrelated events. A damp patch, linked to the presence of timbers, with a shower tray on the floor above, all taken together may suggest the cause and the possible consequence of the failure.

STANDARD PARAGRAPHS AND CLAUSES

There are certain paragraphs that surveyors tend to use regularly – where these inform the client this is acceptable. It is important that guidance upon the consequence of a discovery of, for example, subsidence, asbestos or dry rot be explained in precise and accurate terms. It is important, however, that the survey report deals with the problems that your client will have to resolve if they were to acquire this property.

Standard phrases enable the surveyor to attain a uniform standard of report and make sure that there are no minor variations in the information contained within these paragraphs. However, it is essential that the report is the product of considered opinion and not a selection of alternative paragraphs.

Words used as illustrations

Many elements of the building have technical terms that are not recognized by all clients. A sketch of a property with the various technical words connected to their location can make life easier. The use of photographs greatly aids the appearance of a report, making it more attractive and easier to read and understand.

CLIENT EXPECTATION

A full explanation of what will be carried out, what will be done, and what will not be done must be set out in pre-contract discussions.

Mortgage valuations

In 1980 these limited reports were released to the public who were paying the building society for them. This may have reinforced the public perception that the valuation was providing all the information required about the condition of the building. This misunderstanding has never been successfully contradicted by the building societies. As a result of the litigation that followed public unhappiness with these reports, the status of the mortgage valuation has been enhanced. This minimal inspection has come to have a liability out of proportion to the task.

The valuer is supposed to follow the trail of suspicion irrespective of the cost to discover the extent of any deficiency. The Construction Industry Council (CIC) definition of a mortgage valuation is that it is to provide an opinion of the price a property might achieve if it were sold, or of the rent if it were let. A valuation should not be interpreted as, or used in substitution for, a survey report.

These 20-minute inspections are intended to identify readily visible defects, not to be a detailed record of the absence of deficiencies in various parts of the building. In order to say that no defect existed there is a need for a more detailed record of what was there at the time of the valuation.

It is unreasonable to expect a detailed consideration of the condition of the building in a mortgage valuation.

EXAMPLES OF SOME STANDARD CLAUSES

Assumptions

The surveyor will assume the following in preparing this report (and valuation):

* No deleterious or hazardous materials or techniques have been used and that the land is not contaminated; or
* No high alumina cement concrete or calcium chloride additive or other potentially damaging material was used in the construction of the property or has since been incorporated.
* There are no unusual or especially onerous restrictions, encumbrances or outgoings which may affect the property.
* The property has good title.
* The value of the property is not affected by any matters which would be revealed by a Local Search, Replies to Enquiries before Contract, or by a Statutory Notice, and that neither the property, nor its condition, use or proposed use, is, or will be, unlawful.
* The surveyor will open trap doors and hatches where they are visible, accessible, free to be opened and can be opened safely and easily.
* The surveyor will neither inspect nor report upon any areas which are covered or unexposed, irrespective of the method of fixing of the covering that may exist.
* We strongly recommend that you read all the report and then consider, with our help if you wish, the wisest course of action. The condition of the property and the risk of future expenditure are usually relevant to any decision to purchase and what price to pay.
* We have assumed in compiling this report that all necessary building regulations approvals and consents were obtained and complied with when the alterations were carried out to ...

The following points should be checked with your solicitor to ensure retention of any rights or guarantees which should be reserved for you and to clarify any liabilities you may have to others.

(a) The ownership of the perimeter fences and party walls.

(b) Any responsibilities for the maintenance and upkeep of the jointly used gutters, rainwater downpipes, drains and chimneys.

(c) Any responsibilities to maintain the side and rear access alleyway which we believe is private and not adopted by the local authority.

(d) The presence of any certificates of guarantee and accompanying reports and plans in general, and in particular for the double-glazing, damp proof course, timber treatment, or electrical installation.

(e) Obtain copies of the relevant planning and building regulations approvals relating to the renovation scheme and/or extension.

(f) Clarify the right-of-way in favour of this property and any others over the side and rear access ways.

(g) Establish if there are any outstanding credit agreements. (These may relate to kitchen fittings, double-glazing, central heating or similar retro fit.)

(h) Establish if there is, or has been, a boiler maintenance contract.

Liability limited to addressee only

The reports shall be for the benefit of the addressee only. We accept no liability to any other party who may seek to rely upon the whole, or any part, of this report.

Surveys of Specialist Buildings

THE SURVEY OF TIMBER-FRAMED PROPERTY

CHECKLIST

EXTERIOR

If in doubt, examine the following areas of the building to assist with the diagnosis.

- Provision should have been made to allow the timber frame to shrink and the slight expansion in the brickwork to have taken place. Is there a gap between the head of the outer brick skin and the fascia soffit?
- Weep holes. These may be located below the damp course level in brick external skin construction. In cavity wall construction this would not be the case.
- Windows. The windows will be fixed to the inner leaf and not the outer leaf of brickwork. There is therefore a deeper reveal than is common with brick and block construction.
- Movement. There should be signs of provision for movement around windows and doors. The gap would allow for the shrinkage of the timber frame.

ROOF VOID

- Is it ventilated? (3,000mm^2/m^2) All roofs should be ventilated to this standard. A credit card is 4,500mm^2, so one is looking for ventilation at a rate of a credit-card sized gap per 1.5m^2.
- Gable. Examine the construction from inside the roof void. Sometimes timber frame may be used in brick and block construction to form the gable.

* Party wall construction. If the wall is plasterboard faced in the roof void the construction is almost certainly going to be timber frame.

INVESTIGATION

* Electrical switch or socket installation. Loosen a cover plate. In some cases one may be able to see the construction into which the switch has been placed.
* Under the sink. Most work is of a lower standard in this location and the construction may be visible at pipe or waste entry.

WALLS

* Wall construction. Tap the walls. They will sound hollow. As most modern construction uses dry lining that will not identify the construction. Tap the walls over windows. In timber frame there should be a more solid sound because of the lintels placed at this position. No lintel is required in the support frame to the dry lining.

CONSTRUCTION REQUIREMENTS

* Cavity barriers: vertical. Vertical fire barriers are to be installed if the length of the cavity exceeds 15m (Scotland), 8m (Northern Ireland), at partition walls (England and Wales).
* Cavity barriers: horizontal at ceiling levels. These should be protected with a dpc tray with the breather membrane lapped over the tray. Fire barriers are inserted at each floor level.
* Cavity barriers: openings. Scotland and Northern Ireland – around all openings; no requirement in England and Wales.
* Cavity barriers: chimneys. To the outside of the frame and around chimneys (25mm).
* Fire stops at party walls. Vertical barriers are located in the cavity to each face of the party wall, at the junction of separating walls, and the roof and compartment walls and compartment floors.
* Membrane: breather. The breather membrane, set against the cavity, should be

self-extinguishing, water resistant, and strong enough when wet to resist site damage. The most common material used in Tyvek.

+ Membrane: vapour barrier. This is placed on the inner face of the wall construction. The aim is that it should restrict the passage of water from the building into the wall construction. This is achieved with 500 gauge polyethylene sheet or vapour control plasterboard. Foil backed plasterboard is not acceptable.

+ Wall ties. The ties must be able to allow for differential movement. The maximum vertical spacing is 450mm and 600mm horizontally. At the jambs of openings the vertical spacing is 300mm.

+ Holding-down bolts. These must be of a durable material. The sole plate should be anchored. Any anchorages to the panels shall be fixed to the vertical studs.

THE SURVEY OF RESIDENTIAL FLATS

The survey of a flat levies a greater challenge upon the surveyor. Not only has the surveyor to identify the condition of the flat, but also they must decide upon the implications of the condition of the remainder of the building within which the flat may be located.

There are three functions to such a survey. The first will be to examine the flat and all those parts of the building that would directly impact upon the occupant of the flat. The second function of the inspection will be to consider the quality of the communal services of the building which the occupant of the flat will make use of. The third function of the inspection will be to consider the liability that the remainder of the building will impose upon the occupant of the flat.

The RICS recommends that the surveyor must

endeavour to establish the extent and nature of the repairing liabilities under the lease, the client's potential responsibility for executing repairs and also the liability to pay for repairs executed by others.

The surveyor must either examine the lease and comment upon its implications or, if the lease is not available or is too complex for the liabilities to be identified and understood on a first reading, express clearly the limitations of the surveyor's advice.

THE LEASE

The lease will contain a number of key covenants.

THE LESSEE'S COVENANTS

The demise – this sets out that part of the property for which the lessee will be responsible, have access to or have the right to pass over. It should also identify the area for which the landlord is responsible to repair.

Access – the landlord has no right of access to a property that has been leased to a tenant unless such a provision is included within the lease.

Re-entry provisions – many commercial leases include a covenant which gives the landlord the right to enter onto the property if the tenant has failed to comply with a notice to repair. The time that must elapse from the service of a notice to the landlord's entry is usually stated.

Schedule of condition – some leases have a reference to the repairing obligation being limited, whereby the lessee shall have no obligation to return the property in any better state than that set down in a schedule of condition. Firstly, make sure that such a schedule is attached to the lease. Secondly, hope that the schedule is worded in such a way that it can be applied to limit those areas that are in question.

Statutory requirements

Most leases require the lessee to undertake all works that may be required by an Act of Parliament, by a local bye law or similar occurrence. This covenant can require the lessee to upgrade, improve, or amend the demised premises if the law should change and require alterations to residential property (or in the event of a commercial lease, commercial property). The schedule below gives some idea of the ordinances that may influence the liability of a leaseholder.

i. Building regulations – all works will have to comply with the requirement of the building inspector.

ii. Means of escape – the accommodation must comply with the existing fire certificate, or any amendments that may be issued during the lease.

iii. Public health – this relates to the condition of kitchens cooking food for public consumption (such as in a hotel, pub or restaurant). This will also control the method of removal of contaminated or deleterious materials.

iv. Town planning – covers the listing of buildings or the creation of conservation areas. May require specific roof coverings, or control the type of windows that are allowed to be inserted.

Use

This sets out what may happen in the demised premises.

Alterations

This covenant controls the work that the lessee has a right to undertake to the property. It sets out the method of obtaining a licence for alterations and identifies what the landlord may resist, and whether changes have to be reinstated at the term end.

Repairing covenants

The obligation to repair will relate to the demised premises alone. The repairing obligation will usually be expressed as being 'to keep the premises in repair'.

Yield up

This sets out the state and condition of the building at the end of the term, and how it is to be handed back to the landlord.

Fee recovery

The landlord is unable to recover the cost of those fees that may have been incurred in preparing a dilapidation claim unless the lease contains a covenant that entitles the landlord to recover those costs.

Decoration

This covenant will set down the frequency for internal and external decoration and the quality and extent of the work that may be required.

Leases vary, and these notes are to assist the surveyor in looking at key parts of the lease.

Service charges

An examination of the lease will probably reveal that the lessee has to pay a proportion of the outgoings on the main buildings. If that is more than 15% of the total expenditure advice should be provided on the complete property. The surveyor should also comment upon the implication of the landlord's covenants. There remain some rogue leases where neither party has covenanted to carry out important repairs, or the service payment is a fixed and inadequate amount.

Attempt to obtain the service charge accounts for the previous three or four years. It may be necessary to seek specialist advice about the services installation, or the tax position created by some future funding of remedial work. If the accounts are not available it may suggest poor management.

The major expenditure in the running and maintenance of mansion blocks of flats fall into a number of categories:

* roof;
* external decorations;
* lift maintenance;
* communal heating and hot water;
* staff and staff accommodation; and
* internal decorations.

Arrangements should be made to inspect concealed parts of the roof, the lift motor room, the communal boiler and central heating plant, so that one may report upon their condition. Where the maintenance liability is substantial it may be prudent for the services to be inspected by a services engineer.

Lifts

It is a requirement that a competent person inspects lifts twice a year. That inspection will include:

- Electrical safety.
- Safety devices, cables and brakes.
- Wear in moving parts.
- Completion of certificate.

These reports should be available and include a schedule of the maintenance that is required. Those requirements should be followed through to confirm that they were completed.

- The lift cables should be lightly greased and there should be no indications of any irregularities over the surface such as minor strands of wire breaking away.
- The lift should be smooth in operation and should stop accurately at each level.
- There should be no springing or jerking when the car stops, as this may indicate slackness in the cables.
- In the vertical ascent there should be no bumps as the lift touches one or other sides of the shaft.
- The lift should respond easily to the touch of the call buttons and it should not be necessary to press any button twice.
- The alarm system within the lift cage should be tested, after consultation with the caretaker. The absence of an alarm button should be noted within one's report.
- There should be no gaps in excess of 6mm around doors.

An examination of the control room:

- During the operation of the lift there should be no arcing between the cut-outs on the board.
- The room should be clear of rubbish.
- There should be a record of service inspections.

Access for the disabled

The communal areas of residential accommodation will probably have to comply with the requirements of the Disability Discrimination Act 1995 by 2004 if the landlord is regarded as the service provider.

Communal heating system

If the building has a communal heating system that provides central heating and hot water for the flats and the common parts, it is probable that this is going to be a major cost in the service charge. Communal heating is uneconomic and is more expensive than the cost of operating independent heating systems. Comment upon the following within your report:

- Fuel used.
- Boiler type and efficiency achieved.

* Age of life expectancy of equipment.
* Adequacy of ventilation provision.
* Clipping of pipework.
* Mixed materials used in pipework.
* Accessibility of drain off valves.
* Presence of any leaks.
* Materials used in insulation of visible pipes and any risk of asbestos.
* Quality of boiler controls.
* Presence of earth bonding.
* Existence of any pipes touching.
* Condition of calorifier.
* Winter/summer boiler provisions.

Caretaker's accommodation

The service charge will often include the cost of the notional rent and the maintenance and repair of the interior of the staff flats. As this has a direct bearing on the expenditure which the client will face, their inspection will enable the surveyor to advise upon the future maintenance charges relating to these flats.

Means of escape in case of fire

There should be a half-hour separation between the sources of fire risk and the exit lobby within the accommodation, and a further separation of half an hour between that lobby and the communal protected escape route. The internal doors to kitchen, sitting rooms and any room with a boiler should all be half-hour fireproof and self-closing. Complex buildings may incorporate specialized escape systems. These may include the roof top bridges and external fire escapes. The client's solicitor should check with the local authority to see if the building has the relevant and current fire certificates.

If the escape depends upon a secondary means of escape, make sure that you have examined the condition of the adjoining premises which form part of the escape route. The means of escape over the adjoining premises must also be in good order and the access onto the adjoining premises be safe and secure.

Fire protection
Are there adequate means of escape within the building? For property divided into flats, office units or commercial units, there is a need for adequate means of escape to be provided. The main staircase will be a protected route, and each unit would have a lobby arrangement between the unit of accommodation and

the protected staircase. The division between these two areas of the building would have to be one-hour fire protected. Buildings with floors above 22.5m will require secondary escape routes.

Primary escape route
Access to the main escape route should be self-closing fire-check doors. They should close on their own. If they do not, there is a risk that a fire, or more smoke, will spread. What is the condition of the self-closing mechanism?

Alternative escape routes
For tall and wide buildings there will be a need for alternative methods of escape. This may be by way of an alternative staircase. This could be on the inside or the outside. It is important that this staircase is maintained in good condition. Any storage of material within an escape route should be referred to in the surveyor's report.

THE SURVEY OF COMMERCIAL PROPERTY

The report on commercial property must relate to the suitability of the property for the client's needs. Some parts of the structure may need specialist testing, for example, a concrete frame containing high alumina cement. The consent of the occupant will have to be obtained before undertaking destructive testing.

Services

The service installations of the property will form a separate part of the report. This will deal with the statutory services to the property and will record the condition of the equipment, as well as commenting upon its future life.

The services may be listed as follows:

+ electrical services;
+ hot and cold water plumbing installation;
+ heating services;
+ air circulation or air conditioning facilities;
+ drainage services;
+ window cleaning equipment;
+ fire alarm system;
+ intruder alarm system;
+ telephone system;
+ escalators

* hoists/lifts; and
* moving platforms.

Consider:

* cost-in-use;
* projected maintenance costs
* the need and access for cleaning the exterior;
* the risk of vandalism;
* consents that should exist for use or emissions;
* the legislative requirements; and
* provide a competent and informative summary to each part of the report.

LEGISLATION

Town and Country Planning Acts

These Acts control the building's use.

A1 Shops – including post offices, travel agents, hairdressers, funeral directors and dry cleaners.

A2 Financial and professional services

A3 Food and drink

B1 Business – offices, research and development, light industry appropriate in a residential area.

B2 General industrial

B8 Storage and distribution – including open-air storage.

C1 Hotels – including boarding and guesthouses where no significant element of care is provided.

C2 Residential institutions – including boarding schools, residential colleges and training centres.

C3 Dwelling houses – including houses occupied by up to six residents living together as a single household, including a household where care is provided for residents.

D1 Non-residential institutions – surgeries, nurseries, day centres, schools, art galleries, museums, libraries, halls and churches.

D2 Assembly and leisure – but not including motor sports, or where firearms are used.

Applications for planning permission are always required for material changes of use involving amusement centres, theatres, scrap yards, petrol filling stations, car showrooms, taxi and car hire businesses and hostels.

The government is consulting on changes to these Use Classes Orders in 2002.

Building regulations

They set out the method of building work that is acceptable. There is a short period (six months) for investigation after the completion of the building, unless faults occur. Alterations may result in non-complying works having to be upgraded.

Fire Precautions Act 1971 etc.

Certificates required for buildings in multiple occupation. The conditions compliance should be verified.

Offices, Shops and Railway Premises Act 1963

Much of its content is now incorporated in the Workplace Health, Safety and Welfare Regulations. Premises should have:

* a minimum office temperature of 16°C within an hour of work starting:
* sufficient lighting in offices, corridors and stairs;
* an adequate supply of fresh or purified air, supply of drinking water and arrangements for hanging and drying of clothing;
* stairs kept in good order, with suitable handrails; and
* guards on dangerous parts of office machinery.

The Workplace Health, Safety and Welfare Regulations 1992

These regulations set out the minimum safety and health regulations for the workplace. They apply to any non-domestic place of work, and will apply to staircases, lobbies, rooms, roads (other than public) or other places to which the worker would have access at the place of work.

* Emergency lighting must be provided in any room within which the occupants would be in any danger in the event of the failure of artificial light. Category 2 lighting is required where the room is dedicated to work on computer screens.
* Space provision per worker should be $11m^3$. This is calculated by allowing a maximum room height of

3m. That tends to mean that each worker should have a minimum space equivalent to 4m^2.

* Fencing shall be provided in any place where a person could fall 2m or more.
* Toilet provision shall be one for the first five employees, and then one more for every 25 employees above five.
* An adequate supply of drinking water shall be available and cups must be provided.
* In multi-storey property, safe access to the exterior of the building must be provided for the cleaning of the windows (if the windows are to be cleaned).
* The likely cost involved in complying with the terms and conditions of this Act must be referred to in the report.

The Environment Act 1995

Imposes liability upon the polluter. The cost of clean-up operations will only be incurred by the owner or occupier if the contamination is causing unacceptable hazards to the environment or public health.

Disability Discrimination Act 1995

The Act makes it unlawful to discriminate against the disabled on the grounds of their disability and requires service providers to have, after 2004, a non-discriminatory environment.

CHECKLIST OF CONTENTS OF A COMMERCIAL SURVEY REPORT

THE BUILDING

Condition and existing defects

Cost of repairs

Risk of future problems

Deleterious materials

Adjoining buildings

Noise

Summary and recommendations

SERVICES

IT – the provision of ducts and cable ways for IT installations

Electricity – including lighting provisions and their compliance for new space use

Gas

Hot and cold water plumbing

Drainage

Heating

Air circulation/air conditioning

Lifts

Escalators

Hoists/moving platforms

Waste and drainage

Window cleaning

Fire alarm, emergency lighting and means of escape

Intruder alarm

Telephone

Communication equipment

Summary and recommendations

LEGISLATIVE AND OTHER CONTROLS

Town and Country Planning Acts

Building Regulations

Fire Precautions Act 1971

The Workplace Health, Safety and Welfare Regulations 1992

Health and Safety at Work etc. Act

The Environment Act 1995

Disability Discrimination Act 1995

Licences/controls

COST IN USE

Immediate expenditure on repairs

Annual expenditure in use

Planned maintenance

Future alterations and repairs

Expenditure required to comply with requirements beyond control of client

Energy conservation

Insurance assessments

Vandalism

DILAPIDATIONS – LEASEHOLD PROPERTY

The implication of the covenants of the lease

Liability at lease end

Replacements that will be required during currency of the lease

Obligation to decorate and the intervals required

Obligations to insure, to obtain certificates at set intervals

Additional cost of any alterations required in order that they comply with new legislation

Is the building free from design and construction problems?

CHECKLIST OF THE HEALTH CONSIDERATIONS IN COMMERCIAL BUILDINGS

HEALTH IN THE BUILDING

SBS Sick Building Syndrome
If it exists, are there any circumstances where problems would be expected?

ASBESTOS

The surveyor should know where the material has been used and what materials are at risk.

Surveying Equipment

The inspection of buildings is a mainly visual experience, where the eyes do most of the work; the mind directs the eyes, and the hands touch, tap and stroke.

The RICS published a guidance note for surveyors that recommends that certain equipment is used in a building inspection. It is for each individual surveyor to decide upon the appropriate procedure to follow in any professional task. The guidance note applies to England and Wales and lists a minimum list of items necessary:

* Torch.
* Hammer and bolster – for lifting covers or floors.
* A 3m-ladder
* Pocket probe.
* Small mirror.
* Moisture meter.
* Screwdriver.
* Measuring rod or tapes, notebook and writing equipment.
* Plumb line.
* Spirit level.

The Contract

Surveyors owe duties at law to clients and to other people. They need to know what these duties are, how they arise and how best to be equipped to fulfil them.

The formation of a contract does not depend on writing. It is created by an offer and acceptance between parties who intend to enter into a legally binding relationship. A contract may be founded on words spoken on one occasion, or on conversations spread over a period of time. It may be implied from conduct.

No binding contract exists when there is no intention to create legal relations or when a party has not furnished valuable consideration for the promise that they seek to enforce. This valuable consideration can vary from a peppercorn to large sums of money.

Common law has largely left it to the parties involved to conclude the terms on which they contract, and rarely interferes with the relationship. As an exception to that general rule, it implies terms where the contract is silent but which the parties would regard as too obvious to merit express inclusion. Common law can in certain cases override a contract and render it void. Where a contract offends against public policy, for example, by being founded on illegality, it will not be enforced.

TERMS OF ENGAGEMENT

The confirmation of the instructions letter should be copied into the report and included as an appendix. The following information should for be included.

CONFIRMATION OF INSTRUCTIONS

The address of the property, including:

* brief description of the building that is to be inspected (based upon information provided).

The client and their address.

Date of:

* Instruction.
* Proposed inspection date (AM or PM).
* Proposed date the report to be issued.
* Date of confirmation of instruction to be received.
* Cancellation costs, date/time up to which cancellation is accepted.

The cost of the report:

* The fee including VAT for the inspection and report.
* The cost of any options, such as drains, electrical or service testing.
* Hourly rate for any additional work that may be required over and above the commission confirmed.
* Payment required by? For example, prior to release of report/prior to inspection.
* Options for method of payment, e.g. credit card, cheque, etc.

The type of report to be provided. For example:

* Building Survey, HSV, limited inspection, partial report, single issue comment, etc.
* Sample pages of a typical report of the type commissioned.
* Guidance upon report format.

What you will do. For example:

* Examine all surfaces that are accessible and exposed.
* Test for dampness the internal wall surfaces to ground floor at skirting level where accessible.
* Lift floor coverings where possible to examine the floor surface in not more than three random locations.

What you will not do. For example:

* Move heavy furniture.
* Lift Carpets.
* Test, e.g. drains, electrical installation and services.

Detail of any options that you offer. For example:

* Drains test at an approximate cost of £ ....
* Electrical test at an approximate cost of £ ....
* Service testing at an approximate cost of £ ....

The limitations of your inspection. For example:

* Surfaces will be viewed from within the boundaries of the property, and may use binoculars and a 3m-ladder.
* Lifting nailed carpeting for limited random inspections.

The report contents

* Advice upon the defects that have been discovered.
* Advice upon the risk of any defects suspected within concealed parts of the property.
* Advice upon a budget guesstimate of the cost of remedy, subject to the customer or client checking the costs before commitment to purchase.
* Recommended repairs, investigations or works to be done before a commitment is made to purchase the building.
* Recommended works that should be done before the property is occupied, but after purchase.
* Recommended works that should be done immediately after purchase, but which could be undertaken within an occupied building.

Requirements from the client

* Notify the surveyor before the inspection of any personal circumstances that may affect the inspection and guidance to be given in the report, e.g. health limitations, proposals to remodel part or all of the building or, specific requirements.
* Confirm the appointment before inspection can go ahead.
* Notify cancellation up to 36 hours before inspection.

The surveyor cannot include in the contract any term purporting to exclude or restrict:

* liability for death or personal injury resulting from negligence; and
* 'business liability' for negligence, except insofar as the term satisfies 'the requirement of reasonableness' (Unfair Contract Terms Act 1977).

EXEMPTION CLAUSES

Any contractual term which has not been individually negotiated is to be regarded as unfair if, contrary to the requirement of good faith, it causes a significant imbalance in the parties' rights and obligations arising under the contract, to the detriment of the consumer.

The Supply of Goods and Services Act 1982 implies in all contracts for the supply of services terms that the supplier will carry out the service with reasonable care and skill and within a reasonable time (unless the time for performance is provided for by the contract or the course of dealing between the parties).

By s.1 of the Contracts (Rights of Third Parties) Act 1999, a third party has the right to enforce a term in a contract made between others unless it appears that the parties did not intend the term to be enforceable by the third party.

CHECK LIST OF CONTRACT AT TIME OF INSPECTION

- Is the building the same type and size as you contracted to inspect?
- If it is different, does that affect the inspection or the fee that you quoted?
- List all parts of the building to which access is not possible and why.
- List those parts of the building to which access is not permitted at the time of your inspection.
- List those parts of the building that you did not inspect.
- List those parts of the inspection that you undertook to carry out, but which you did not complete.
- List those options that you undertook to provide that were not completed.
- Schedule all variations to the content of the contract that you negotiated.

QUESTIONNAIRE FOR OCCUPANT OF THE BUILDING

* How many years have you owned the property?
* Are you moving far?
* Do you know whether the property suffered bomb damage during the war? Do you know of any bomb damage in the area?
* Have you made any alterations to the building? If so, where, and was the work carried out under a guarantee? May I see the guarantee?
* Have you had any major repairs carried out during your ownership?
* When was the house last rewired or the wiring tested?
* When was the plumbing last altered or renewed?
* When were the drains last tested or altered?
* Have you had any particular problems with dampness inside the house, in walls or floors?
* Are there any existing guarantees for double-glazing, damp courses, rewiring, heating controls, boilers or other parts of the building, and may I see them?

The Duties of Surveyors, their Performance and their Breach

The standard of care which is required of a surveyor or valuer carrying out an inspection of a property should be no less than that required of the average practitioner within the profession.

1. CONSTRUCTION

The surveyor must identify the building construction that has been used in the house or other property under inspection. They must be cognizant of the history of that type of construction and be aware of specific failures that have been found in similar properties.

2. OMISSION

The surveyor is responsible for carefully inspecting the property and reporting upon the defects that are evident to them. In the event of defects which were in existence and visible at the time of the inspection not having been reported, they will be negligent provided that it was reasonably practical to have noted, recorded or seen the failures in existence. It is unreasonable for a surveyor to be expected to find defects that are concealed by the building's construction unless there was evidence, which suggested that there was a risk of failure in concealed parts of the building.

3. PROJECTION

The surveyor is responsible not only for recording defects that were evident but also for advising on the likely consequences of those failures noted.

THE PRINCIPLE OF DUTY

The quality of the surveyor's work will affect others, and if they fall into error both their client as well as 'other parties' may be able to make a claim.

Apart from contract, no action for negligence will lie unless the claimant can show that the defendant owed them a duty of care at the time of the act or omission of which they complain.

The invitation to rely, and the resulting reliance, frequently will result in a contract between surveyor and a client. Such a contract contains an implied term that the surveyor will carry out the contractual work with the skill and care reasonably to be expected of a competent surveyor.

An action in tort will not be open to one contracting party against another if the terms of the contract specifically preclude it (*Henderson* v *Merrett Syndicates*).

Successful third parties' actions on their reliance on surveys and the resulting reports include:

* The building owner (not to damage the property or to cause them personal injury).
* A 'close and direct' relationship to whom the surveyor knows that the report will be shown (*Hedley Byrne & Co. Ltd* v *Heller & Partners Ltd*).
* A prospective purchaser of the house specifically in connection with a particular transaction (whether advising the vendor or the representative, bank or mortgage supplier of the purchaser).

Where it is reasonably foreseeable that a person may suffer personal injury or damage to property in consequence of an error in a report.

* The client whom the surveyor ought to have warned that his garage roof was made of asbestos, climbing on it would be most unwise (*Allen* v *Ellis & Co.*).
* A careless mis-statement that causes physical injury or damage to property other than the property that is the subject of the statement (*Clayton* v *Woodman & Son (Builders) Ltd*; and (for the qualification) *Murphy* v *Brentwood DC*).
* The claimant unconnected with the client who is injured by the failure of, for example, guttering that the surveyor carelessly failed to notice is insecurely fixed and might fall to the highway and kill or injure whoever is unfortunate enough to be passing.

• When carrying out a survey on behalf of a client who is not the occupier and a feature likely to cause injury to the occupier is observed, the surveyor should inform the occupier as well as the client. If the surveyor fails to do so, they could well be vulnerable to the suggestion that by virtue of their skill, their presence in the occupier's home, and the danger that has been observed, they owed the occupier a duty to take reasonable care.

Specialists will be judged by the reasonable standards of the speciality in which they hold themselves out as being skilled. Yet even where professionals hold themselves out as having especially high professional standards, the standard to be applied is that of the ordinarily competent practitioner of the particular profession, as expressed by Bristow J:

> The duty of a practitioner of any professional skill which he undertakes to perform … is to see the things that the average skilled professional in the field would see, draw from what he sees the conclusions that the average skilled professional would draw, and take the action that the average skilled professional would take. (*Daisley* v *B. S. Hall & Co.*)

Allowance is also made for the fact that on a particular matter professionals may belong to one of two or more reasonable schools of thought, and for the fact that some professionals are by temperament or training more cautious than others.

Performance and breach of duty: some examples.

(i) Failure to carry out instructions.
(ii) Inadequate knowledge or experience for the work undertaken.
(iii) Inadequate inspection.

In some cases a surveyor has been negligent in omitting from the inspection:

• a part of the building, e.g. the rafters of an old cottage (*Hill* v *Debenham, Tewson & Chinnocks*);
• the roof spaces of a manor house (*Stewart* v *H. A. Brechin & Co.*);
• the cellar of a house (*Conn* v *Munday*); and
• the structure of the building of which the flat forms part (*Drinnan* v *C. W. Ingram & Sons*).

A failure to uncover and open up

* A failure to make proper use of instruments may lead to a finding of negligence.
* Failure to observe. The surveyor must be careful not to miss important evidence to be derived from, for example, trees adjacent to the survey property.

(iv) Inadequate report

Ideally, surveyors need to be masters of English prose and must be careful of style as well as content.

* Do not use words of commendation so broad that they cannot be justified.
* Warnings should be well expressed and cover all the matters that call for warning.
* Dry rot calls for a warning if active, it should be reported on even though dormant.
* All risks should be evaluated carefully and clearly. If a court took the view that a particular part of the building could and should be inspected and reported on, a surveyor will not be protected by a statement that this part was not inspected.
* The surveyor cannot, at the stage of writing the report, introduce into the contract terms not contained at its formation.
* An allegation of negligence cannot be rebutted simply by reciting places that have not been examined. The surveyor should make clear to the client the extent to which, by reason of the limitations of the examination, reliance on the report ought to be withheld.
* The surveyor should not only evaluate the risk associated with features that they have seen. On the basis of what has been seen, the surveyor should try to evaluate for the client the risk of defects lying in the parts that have not been seen.

CAUSATION AND REMOTENESS

Once a claimant has established a breach of duty on the part of the surveyor they must identify the difference that the breach caused. Some part of the loss and damage that they have suffered may remain uncompensated for on the basis that it is too remote.

(i) Causation

The claimant will have to show that they relied on the surveyor's report and that this reliance was an effective cause of the train of events that followed. A claim can be defeated or limited if the claimant effectively owes their position to some cause or causes other than the surveyor's report.

(ii) Remoteness

A defendant need not pay for a consequence that could not reasonably have been foreseen at the time when the breach of duty was committed.

Heads of damage common to all claims

* Purchasers may show that they would have refrained from buying the property concerned and that they would have been saved a number of consequences. They may recover expenses associated with any repairs not advised to be required, such as the cost of accommodation for people and storage for furniture.
* The claimant in a given case may recover losses incurred by virtue of repairs, e.g. the temporary loss of the value of the use of the property, whether this is expressed in a loss of rent or the loss is of a general kind.
* A purchaser who decides to sell the defective house may recover the expense of two removals within a short time. They should recover the difference between their actual expenses and the expenses that they would have incurred had they made only one move after receiving the surveyor's report.
* Physical inconvenience and discomfort. On several occasions they have recovered general damages in respect of physical inconvenience and discomfort. Damages may cover mental distress and frustration, provided that they are caused by physical discomfort or inconvenience resulting from the surveyor's breach of duty.
* The court has power to award interest on sums awarded in proceedings. The rate of interest and the period for which it is to run are at the court's discretion.

Contributory negligence

The damages recoverable by claimants in negligence may be reduced to take account of the extent to which their own negligence contributed to the loss and damage that they suffered.

Limitation
In the majority of cases six years after it arose. In the case of an action founded on personal injuries or death (as opposed to one in which a personal injury claim is ancillary to a claim in respect of financial loss) the relevant period is three years.

Health in Buildings

The surveyor should be aware of those areas of building construction that may create health risks and advise a purchase of those risks.

'Sick Building Syndrome' (SBS), if it exists, is an emotive description of conditions that are believed to have an adverse effect upon the people who work within a building.

When investigating a complaint check the following:

* sickness records;
* record of use of medicine cabinet to note paracetamol use; and
* the amount of coffee drunk.

The 'symptoms' of SBS include the spread of infection within a building; above average absenteeism; tummy upsets or frequent headaches among the employees; or injurious circumstances that can lead to serious illness or even death. It is not only office buildings that may be affected. Current trends to increasing insulation and reducing air-changes will create more problems in the future.

CAUSES OF SICK BUILDING SYNDROME

Up to 30% of refurbished buildings and an unknown number of new buildings may suffer from SBS. Extensive research has shown that there are a number of features that are associated with the symptoms of SBS. These include:

* a hermetically sealed, airtight building shell;
* mechanical heating, ventilation and air conditioning systems;
* use of materials and equipment that give off a variety of irritating and sometimes toxic fumes and dust;
* unsuitable lighting;
* application of energy conservation measures; and
* lack of individual control over environmental conditions.

Water-borne bacteria

Common design features include:

* Drinking water supply.
* Commercial properties.
* Contamination of drinking water.

In public and commercial buildings notices are to be fixed to all water supplies in toilets and washrooms showing if the water is suitable for drinking.

Residential properties

* The water tank must have a secure cover.
* Water tanks should always be kept clean.
* The tanks location must also be kept clean and free from birds, vermin, etc.

Breeding grounds for bacteria

* Leather or rubber washers on taps.
* Mixer tap filters on the spout.
* The build-up of calcium in a showerhead.
* Hemp: used as a jointing seal on metal pipes.
* Boss white or plumber's metal.
* Ferrous oxide: in rust or sludge.
* Directional spouts on the kitchen tap.
* Filters: on commercial premises.

Legionella bacteria
Legionella pneumophila bacterium is found in the mains water supply, but in a concentration that does not present any risk. It thrives in the temperature ranges of 25–35°C. Below 10–15°C it survives, between 20–45°C it can multiply, but above 55°C it is killed. The water temperature ranges are found within domestic and commercial property. The summer temperature of the water in a cold-water storage tank can be sufficient for the propagation of the bacteria. Simple ingestion by drinking does not cause infection.

Contaminated water in water-cooled air conditioning systems has been a prime cause of infection. These are now being phased out, but some still remain in older buildings. Inhalation is possible through showers, one of the earliest causes of fatalities from this disease. The water agitated in whirlpool baths, or in taps, can provide air-borne particles that can be inhaled

Prevention – avoid long pipe runs and any stagnant water serving shower or bath taps.

Sundry sources of contamination

+ Vacuum cleaners.
+ ISP. Indoor Surface Pollution. This can be remedied by the use of liquid nitrogen or hot water extraction cleaning (steam cleaning).
+ Fibres. Dust can be injurious to health.
+ Asbestos. Recognition is difficult.
+ Chemicals. Wood preservation, nitrates leached from farmed land, cavity fills, urea formaldehyde, and glues in construction.
+ Water supplies. Lead mains water pipes and nitrates from fertilizers.
+ Site infill. Toxic waste products used in the landfill.

RADON GAS

Most of the UK has a very low level of background radioactivity emitting from the ground. Some exposure to radon is unavoidable since the gas is given off by all earth materials. Minute quantities are always present in the air. Radon levels are higher indoors than outdoors because the gas is able to accumulate in confined spaces. The concentration in homes is influenced by the geological nature of the site and the structural detail of the building. The levels of radon will vary with the seasons and during the day. The levels within buildings in the summer tend to be less because more windows are open and there is a greater level of ventilation. The reverse is true in the winter.

Radon levels will depend upon:

+ weather conditions for the entire year;
+ geographic location;
+ floor construction;
+ punctures or holes in floor;
+ age of construction; and
+ extent of ventilation.

ASBESTOS

Asbestos is a generic term for the fibrous forms of several mineral silicates that occur naturally in the seams of many igneous or metamorphic rocks. The health risk associated with inhaling asbestos fibres was not established until 1920.

Composition of asbestos

There are six types of mineral which are commonly called asbestos:

* chrysotile (white asbestos);
* crocidolite (blue asbestos);
* amosite (brown asbestos);
* anthophyllite;
* tremolite; and
* actinolite.

Crocidolite, amosite and chrysotile are the only forms of asbestos used to any significant extent in buildings in the UK. Whilst current regulatory standards differentiate between the health effects of blue, white and brown asbestos, it is reasonable to treat all types of asbestos with equal caution. Care should be taken, because colour is not a dependable guide for identification.

* Chrysotile fibres tend to be whitish in colour and have a smooth silky texture.
* Crocidolite fibres are usually blue and short, being straighter and less silky than chrysotile.
* Amosite tends to be brown with fibres, more brittle than either chrysotile or crocidolite.

Asbestos cement is cement reinforced with asbestos fibre of approximately 10–15% of the total by weight. About 1% of the asbestos cement roofs currently installed in the UK may contain crocidolite.

Roofing felt used to contain asbestos and bitumen. Site operations using these products are likely to generate small amounts of asbestos dust, the production of which could be ignored if small areas are cut in the open.

Asbestos fibres are often added to paint to give it strength, and were used to produce various finishes, such as Artex.

Removal

The Asbestos (Licensing) Regulations impose strict controls on work with asbestos insulation or coating. Only a contractor who holds a licence granted by the Health and Safety Executive may remove asbestos products, unless the work involves short-term repair or maintenance.

Control limits
The control limits for asbestos dust levels in the working environment are laid down by the Health and Safety Executive (HSE, 1995). The UK limits are the most stringent in the world.

Roofs

Most roof defects will only be found if a close inspection is possible. Some surfaces may be concealed and the roof void is not always accessible, making an effective evaluation difficult. Make sure that the extent of the inspection is recorded within the report.

Take account of issues of health and safety before undertaking a roof inspection. The risk of failure in roofs may be increased by:

* Exposure. The location of the roof will determine the design requirements. Building Regulations have been prepared to ensure that roof construction is safe. Where the building is in a location where snow can be expected, the roof should have been designed to deal with the extra weight and penetration. Check for the load bearing capacity of the construction. Think about where snow sliding off a roof may land. Are conservatory roofs at risk?
* Access. Roofs that can be walked upon fail first. Shallow slopes, flat surface or central valley roofs are at risk. Check for damage.
* Rain water discharge. Poor overhangs at eaves increases the risk of water being trapped within the construction. Parapet or central gutters increase the risk of failure.
* Inadequate ventilation. Rapid failure if the voids are not ventilated.
* Structural adequacy. Make sure the roof can carry the surface and any applied loading.
* Non traditional roofs. If the design is not tried and tested, there may be additional risks that have yet to be identified.

FAILURES IN FLAT ROOFS

Asphalt surface crazing

Light surface crazing, no water penetration. Depth will determine risk of water entry. It may have been caused by:

- Variations in surface temperature.
- Absence of effective solar reflective treatment.
- Bonding of asphalt to deck construction, which will mean any deck movement is transmitted into the asphalt and may cause deep splits.
- The asphalt was not treated with sand at the time of its installation, which can lead to bitumen lying on the face of the asphalt, which crazes.

Asphalt surface splits

Failures that run through and allow water penetration. It may have been caused by:

- Solar gain causing continuous expansion and contraction of the surface. Ultraviolet radiation in the atmosphere can cause oxidization that may assist in the breakdown of the surface.
- Movement within the building.
- Varied movement between deck and surrounding walls.
- Deterioration of the deck or structure.
- Ceramic tiles on roof surface trapping water below causing cavitation.
- The application of an epoxide paint, which is supplied as a solar reflective treatment. The paint fully adheres to the asphalt and this strong bond results in the asphalt being damaged by the shrinkage of the paint film.

Provided no crack is greater than 3mm deep they may be treated by the application of a solar reflective surface.

Ponding

Ponding on a flat roof may have been caused by:

- errors in the original construction;
- failure in the roof structure; or
- alterations that affect the discharge of water by the intended fall.

Built-up bituminous felt roofs

The intention is that each of the three layers bond through the use of a hot bitumen which fuses the materials together. Where blisters occur to the surface of the roof this is an indication that the layers of felt have not bonded together to form a single unit. Blisters may be caused by:

- The bitumen not being hot enough when applied. This tends to leave a shiny finish to the bitumen, which does not stick the layers of felt together. Blisters or bumps occur in the top layer.

- Over-cooked bitumen. The bituminous surface beneath the felt is of a charcoal crusty appearance.
- The surface of the felt may have been damp either through rain falling at the time of installation or due to the presence of dew. This tends to cause the bituminous finish to be crazed.
- Inadequate application of bitumen.
- Blisters that are empty and yield to pressure may have been caused by poor workmanship when the surface was applied which has resulted in water becoming trapped to the underside of the felt.

Before deciding upon the condition test the underside for dampness. A concrete deck will take more time to dry than a timber structure, but the timbers may decay as a result of the penetration.

Torch on roof covering

Single layer elastomeric or similar materials require careful application. Joints are welded when the material is laid. Verify the quality of the work, any sign of failure and presence of warranties.

Parapets

Water penetration via a parapet is often mistaken for roof failures. Any small fracture in cement render facing will allow water penetration, which can expand during cold weather. Damp courses are not always used below the coping in a parapet walls, because the damp course will weaken the adhesion of the coping.

Incorrect location of a damp course may result in water penetration to the interior of the premises or into the cavity, from whence water may enter the building. The closing of the cavity by a floor slab requires the installation of an accurate cavity tray. Detailed examination of surrounding areas must be made to see if damage has been caused.

Parapet gutters
The parapet gutter is likely to leak; it is only a matter of when. In part this may be because of the lack of access for maintenance. There is a major risk of failure around the chute that transfers rainwater from the gutter into an external rainwater hopper. Internal pipes, built into the wall, are also at risk of failure. The surveyor must identify the method of water discharge and its condition.

Edge treatments
Examine the roof edge carefully. Weak design or defective construction may allow water entry.

The junction of the wall and roof is prone to failures. In a flat roof inspect carefully the construction and its quality at the roof edge. It is important to carry out a detailed examination of the junction of the flat roof and its discharge into the gutter. In many properties there is an apron flashing to prevent water running back behind the gutter. The flashings should project into the gutter. Where they are cut short water can run behind the gutter and dampen the wall.

Pipes

Failures are common around service pipes passage through the roof surface. Carefully check the sleeve, even a small gap may lead to water entry.

METAL COVERED FLAT ROOFS

All metal roofs have to take into account the high coefficient of linear expansion of metals, which causes substantial surface movement. The metal is laid in small panels. The vulnerability of metal roofs is at the seamed welts or rolls between the panels. It is difficult to examine the seams to ensure that there is a double lap and that the metal is clipped in their length.

Lead

Where lead is used it is vulnerable at all points of curvature because of the thermal movement, which causes stress if the metal is unable to move, for example, where it is fixed or restricted. Sheet lead of the following codes can be recognized by the thickness of the lead. The most common uses are set down in the table below.

Table: Lead

Uses	Code	Thickness (mm)
Soakers, damp proof courses to cavity walls	Code 3	1.32
Soakers, flashings, flat roofing (no traffic)	Code 4	1.8
Flashings and lead slates, roofing, ridges and small gutters	Code 5	2.24
Roofing and gutter lining, flat roof (with traffic)	Code 6	2.64
Roofing and gutter lining, flat roof (with traffic)	Code 7	3.15

PITCHED ROOFS

The pitched roof is usually supported on a timber frame and tiles or slate covers. These may be either tiles of a clay or concrete constitution or thin-layered stones that are usually either slate or limestone. Vegetable matter can also be used as a roof covering and the more common examples of this are thatch and timber shingle.

Weight of roof covering per m^2

- Asbestos cement sheets 16kg.
- Thatch weighs 33kg.
- Slates weigh approximately 42.50kg.
- Clay tiles weigh around 61.50kg.
- Concrete tiles weigh around 85kg.

Whilst examining a roof surface, it is important to consider both the condition of the material and the ease of carrying out any repair. Refer to the extent of surface damage in the survey report. The surveyor should also consider how practical it would be to replace broken slates or tiles.

- Are matching tiles available or are they difficult to obtain?
- What further damage could be caused during the repair?
- Are there any implications created by difficulties of access?
- How easy will it be to erect scaffolding?
- Will access to other property be needed to repair?
- Who is to do the work? A landlord or a tenant, or the freeholder?
- For a slate slope, is it possible to fix a new slate with a tingle?
- How much of the roof will need to be stripped to effect a complete repair?

The surveyor must consider the condition of the existing roof covering and weigh up the practicality of a repair and advise whether the roof requires recovering or if it can be patched. In making this judgment in the report, it is sensible to set out those matters that have been considered in reaching this judgment.

Tile or slate surfaces are fixed onto horizontal timber battens which are laid across the supporting framework of the roof. The tile or slate is either hung or nailed onto the battens. The quality of these fixings and their life expectancy should be established, where possible. If a roof is to be recovered, warn of the need to replace part of the framework supporting the roof surface.

Slates vary in their durability. The blue Welsh slates may be expected to last over 100 years, but some pale and grey-flaked slates are very soft and will fail within 20–30 years.

The nails by which the slates or tiles are fixed deteriorate – known as 'nail sickness'. A slate roof surface, which is over 40 or 50 years old, is liable to suffer continual failures until the surface is stripped off and the slates re-fixed. Modern sheradized nailing has a longer life expectancy. Their corrosion may be speeded up by condensation forming on the back of the tile or slate surface or by water penetration. This may be accelerated by poor ventilation to the rear of the roof covering.

ROOFING TILES

The durability of roof tiles does vary. Some clay tiles are fixed by using a wood or iron peg. In examining a pegged roof make sure that you have checked the condition of these pegs.

Many sand-faced concrete tiles have failed early, some within 30 years. Some have warranties and the manufacturers have been paying for roof recovering.

Many modern tiles interlock, to allow large amounts of water through the tiled surface if the edges have been damaged. Where the original tiled or slated surfaces have been replaced with concrete tiles the roof should have been strengthened.

All low pitch concrete tiles must be laid on sarking felt (or Tyvek) because water will get past the tiled surface. Modern pitched roofs have to be ventilated to ensure that moist air does not become trapped against the timbers of the roof frame. The ventilation is at ridge and within the roof pitch. Air is allowed into the roof at eaves level. Where Tyvek is used as a sarking felt, ventilation is able to take place through the 'felt', which is a breathable membrane. In such a case ventilation does not have to be provided at the ridge or within the roof slope.

A roof failure close to the eaves will result in more water penetration than a failure at the ridge. The replacement of the pointing to the surface of the ridge will often result in the work causing more damage than the benefit of the pointing would merit.

Carefully inspect the roof surface at the junction with the chimneystacks. The rear of the stacks are often finished with a small gutter section which is almost impossible to examine and a regular point of failure.

In many terraced properties the pitched roof is set between party walls, which act as a fire barrier. The junction between the tiled or slated surfaces and the raised party walls has to be adequately flashed to prevent moisture penetration. This flashing must allow a limited amount of movement as the timber roof construction will move as a result of variations in moisture content in the timber and the expansion and contraction caused by thermal variations.

Table: Sizes of slates and their names

Name	Size (mm)	Size (")
Empress	660 × 406	26 × 16
Princess	610 × 356	24 × 14
Duchess	610 × 305	24 × 12
Small Duchess	559 × 305	22 × 12
Marchioness	559 × 254	20 × 10
Wide Countess	508 × 305	20 × 12
Countess	508 × 254	20 × 10
Viscountess	457 × 229	18 × 9
Wide Lady	406 × 254	16 × 10
Lady	406 × 203	16 × 8
Wide Header	356 × 305	14 × 12
Header	356 × 254	14 × 10
Small Lady	356 × 203	14 × 8
Double	330 × 178	13 × 7
Wide Double	305 × 254	12 × 10

COLOUR OF MATERIALS

Natural Welsh Riven Slate	Blue grey
Westmoreland Buttermere	Fine light green
Buttermere	Olive green (coarse grained)
Kentmere	Deep olive
Delabole Random	Vary in colour from rustic red to grey greens

THATCH

The method of thatching varies slightly depending on the material used. The roof must have a pitch of at least 50° and the rafters must over hang the walls with a tilting fillet at the end of the joists. Battens are fixed horizontally across the roof at intervals of about 125mm at the eaves and about 225mm over the remainder of the roof.

Thatching in Norfolk reed

Norfolk reed (*phragmite communis*) is regarded as the finest material for thatching, although the use of fertilizers has reduced the lifespan of the material. It will not bend over the top of the ridge. A separate ridge is laid over the top using a more pliable material, usually sedge. This is often finished with decorative curls and patterns.

Thatching in long straw

The ridge is based on a tight packed roll of reed about 100mm thick. The top sections of reed are folded back on themselves from the roof slope. A smaller ridge capping is then formed. A reed roof is harder edged where the reed ends stand out, whereas a straw roof has a shaggier, smoother finish with the straw lying on the surface. Wire netting is always necessary to protect a long straw roof from birds. It has a life expectancy of up to 15 years. It will probably require re-ridging every ten years and patching on frequent occasions.

Life expectancy
Reed roofs should last 40–50 years and a combed wheat roof will last approximately 25 years. It is difficult to distinguish between Norfolk reed and combed wheat reed as the main difference is in the treatment of the ridge. With the Norfolk reed the ridge would be in sedge whereas combed wheat reed is used for both ridge and roof surfaces. Differences also occur in the treatment around the windows and eaves in a combed wheat reed. In a proper state of repair thatch will never be wet beyond 50mm or so below the roof surface.

THATCH CHECKLIST

The surveyor should make sure that advice is given on:

* extra insurance premiums over conventional roof covering; and
* the greater hazard of fire.

The life span of a thatched roof will be dictated by the roof design. A pitched roof with no valleys or dormers projecting through the thatch will achieve the longest life. If the thatch is of inadequate pitch or there are dormer windows, or roofs draining onto each other, then its life will be reduced. The surveyor must note these factors in advising a client upon the life expectancy of a particular thatched roof.

ROOF PITCHES

Large slates	22°
Ordinary slates	26°
Small slates	33°
Plain tiles	25°
Shingles	26°
Cedar	45°
Oak plain tiles	45°
Thatch	50°

SHINGLES

Shingles are thin slabs of wood used to cover roofs and walls, usually oak but occasionally elm and teak. Modern shingles are cut from western red cedar, which is a very durable lightweight material, which becomes silver grey when exposed to the weather. It shrinks slightly less than other soft woods and is resistant to insect attack. As the shingle's tendency is to curl it is more likely to split if it does bend, therefore, it is nailed twice: through the middle (one batten below the top of the shingle) and just above the end of the shingle in the layer above.

ASBESTOS CEMENT CORRUGATED SHEETING

Mainly used in industrial buildings, the sheets may have a profile that is not available, making repairs difficult. The report must consider the implications of the small amount of asbestos present in the material and the inadequate surface for walking upon.

Sheets are secured usually to the roof frame by hooked bolts or screws. The bolts need washers and a cover to prevent corrosion and deterioration in the metal.

These panels will not support the weight of anyone who walks over them and extreme care must be taken during the inspection. Replacement of sheets usually causes water penetration.

CHIMNEYS

Flues

Check the flue lining to any open fire. If it cannot be seen warn of the risk. It is essential that flues for gas or oil-fired appliances should be lined, because their fumes cause rapid breakdown in the pargeting. Some new homes have been built with purpose built high-sand content flue liners. There have been problems with the durability of these flues where coal or logs have been burnt.

ROOF VOID

The examination of the roof void of a tiled roof should include testing the roof felt for broken tile nibs. They should slide down when the felt is tapped.

In exposed locations the underside of the tiles were pointed to prevent water penetration caused by driving rain. This 'pargetting', as it is called, was carried out in a soft mortar mix usually including lime, and in most places where it is still found there will be failures in the finish. Consideration should be given of any problems that may occur if the pargetting is now in poor condition.

A spray application to the underside of the roof covering is available to 'repair' weak or defective tiling. The durability of the system remains unproved.

CHECKLIST FOR THE EXAMINATION OF THE ROOF VOID

Roof void	Is it ventilated? ($3,000mm^2/m^2$) All roofs should be ventilated to the extent of $3,000mm^2/m$ of roof plan area. A credit card is one and a half times that size, so one is looking for ventilation at a rate of a credit card sized gap per $1.5m^2$
Timber trusses	Are they evenly spaced? It is usual for most timber pre-formed trusses to be positioned at 600mm centres. The timbers should carry marks which confirm the stress grading
Truss verticality	The accuracy of the installation is vital for the durability of the roof. Variations of more than 20mm are critical
Diagonal bracing	An essential component of the modern trussed roof. The diagonal should be at least 90mm X 22mm

Lateral restraint	The metal strap should be fixed to at least three trusses and carried down the face of the internal wall. The final truss must be fitted with noggings at the point of the fixing of the restraint straps. The strap should turn down onto an uncut block (in brick and block cavity gables or party walls)
Water tanks	Main water tanks should be supported on at least three timber trusses
Insulation	The insulation must not block the eaves ventilation. It is not recommended that insulation is carried under the water storage tanks
Roofing felt	Make sure that the sarking felt has been laid with the laps being to the exterior. Where ventilators are cut through the roof slopes, make sure that the felt has been dressed to redirect any water flow around the opening. Tap the felt to hear if any tile nibs are loose. The felt should be supported at eaves level, to prevent water being held in the hollows that will otherwise form at the bottom of the roof slope
Flat roof ventilation	Cross ventilation should be provided over the support timbers. This will require cross battening to support the roof deck. There should be a 50mm clear gap above the insulation between the timber joists
Notching	Not permitted in trussed rafters. No hole to exceed a quarter of the depth of the timber or to be within 50mm of the edge of the timber

Common faults in trussed rafter roofs include:

* Structural stress due to bad storage.
* Manufacturing faults.
* Unauthorized cutting of the timbers.
* Inadequate support for cold water storage tanks.

* Missing diagonal and horizontal bracing.
* Lack of support at the change in the roof line.
* The roof truss not being supported on the wall plate.
* The trusses being fixed out of the vertical.
* Inaccurate alignment of the fixings.

Timber sizes

It is important to check that the timbers used within the roof are adequate. Where the roof relies upon a purlin the clear span is unlikely to be more than 2.5m.

Table: The main span anticipation of various sized purlins

Clear span of purlin for domestic pitched roof			
Purlin size (mm × mm)	Single purlin (m)	Two purlins (m)	Three purlins (m)
50 × 100	0.90	1.10	1.30
50 × 150	1.30	1.60	1.90
50 × 200	1.80	2.10	2.50
75 × 125	1.30	1.60	1.90
75 × 150	1.60	1.90	2.30
75 × 200	2.10	2.50	3.00

Holding down straps
In exposed locations, holding down straps should be used. They are usually of a minimum cross section of 30mm X 2.5mm. They should be fixed at a 2m centre.

Loft Insulation
Check that the loft insulation is 200mm thick and make sure that the cold water storage tank, service pipes and connections are adequately protected. Most roof voids will be inadequately insulated.

EXAMINATION OF GUTTERS AND DOWNPIPES

A failure in any part of this installation will result in water running down the face of walls, and water backing up into the roof spaces and penetrating into the interior of the property. If there has been a failure, the surveyor should comment upon:

* the amount of water penetration;
* the dampness that may have built up within the wall fabric;
* the timbers that may have become damp; and
* the possibility of fungal decay.

Where there are new rainwater downpipes the original pipework was probably faulty before it was replaced. There will be a risk of timber failures having occurred where timber came in contact with the surfaces that became damp before the pipes were replaced.

Consider the design of the drainage. The higher the speed that water discharges from a roof surface, the greater the chance of the water being able to project beyond the gutter. This may be caused by:

* too deep a fascia;
* the use of pantiles, with their troughs and valley; or
* valley gutters.

GUTTERS

Table: The flow capacity of standard gutter profiles

Size of gutter (mm)	Litres/s half-round	Maximum roof surface (m²)	Litres/s nominal half-round	Maximum roof surface (m²)
75	0.38	18	0.27	13
100	0.78	37	0.55	26
115	1.11	43	0.78	36
125	1.37	65	0.96	46
150	2.16	102	1.52	73

The BRE, in Defect Action Sheet DAS 55, points to the absence of porch guttering as a cause of water entry. The small roof will collect water from the vertical wall above the porch, and the roof and wall combined produce a volume of water worthy of a gutter.

Alignment

The alignment of the guttering will affect the flow rate. For example, a right angle bend close to the down pipe will reduce the flow within the gutter by 20%. With metal pipes there should be a space around the gutter and the downpipe to allow it to be painted. A gutter should have a slight fall to the outlet of about 10mm in 3m (1:350). Fascia brackets should not be wider apart than 1m.

What is the condition of the back of the gutter?

This face is rarely painted and is a common location of faults. Ogee cast iron gutters have usually failed.

Consider the implications of water seepage into the building and of the weight of gutter should it fall.

Joints

Are there any indications of leakage from gutter joints? Common indicators of leaks include:

* rust stains around the joints; and
* patterned staining on the fascia board, brickwork or other materials around the joints.

Report if it hasn't rained for some time as this may limit the presence of indications of failure.

Quality of materials

* Has the gutter installation been completed?
* Are there adequate stop-ends to gutter runs?
* If failure of the guttering occurred, what would the natural consequences be?
* Are children likely to be playing beneath heavy cast iron guttering?
* Are cars parked against areas where gutters drop?
* Are there glass roofs which are unprotected lying immediately beneath an old gutter installation?

Capacity

Check the capacity of the downpipe and the area of roof which it is serving. A simple check is set out in the table below.

Table: Drainage capacity

Rainwater downpipe capacity (mm)	Flow rate in litres/s	Guide to the maximum roof area served by a pipe of this dimension (m²)
50 (2″) diameter	2	22 (250 ft²)
75 (3″) diameter	3.5	50 (550 ft²)
100 (4″) diameter	7	95 (1,000 ft²)
125 (5″) diameter	11	150 (1,600 ft²)
150 (6″) diameter	22	210 (2,250 ft²)

Check that:

* The pipes are not blocked.
* Is there any risk of splashes dampening the face of the wall above the damp proof course?

* What are the provisions for cleaning the gulley?
* Where rainwater pipes have been embedded in walls recommend annual maintenance.
* Make sure that rainwater pipes do not have bends.
* What are the potential causes of blockages?
* Do trees overhang the gutters and deposit their leaves during the winter?

Walls

IDENTIFYING FUNCTIONALITY

The functions of a wall may include:

* Structure.
* Enclosure.
* Weather shield.
* Physical protection.
* Thermal barrier.
* Sound barrier.
* Screen.
* Light.
* Decoration.

Thermal performance

For new construction the U value of walls should not be less than 0.35. This requirement may vary depending upon the quality and size of windows provided within the building.

The construction of the wall can take one of four forms.

1. Solid construction, where the mass of the wall supports the building's roof and internal floors.
2. Cavity construction, where one of the two leaves supports the loading and where the other is the external barrier.
3. Frame, where the enclosure is part frame and part cladding – be it placed on or within the frame.
4. Combination.

MATERIALS

The walls may have been constructed from a range of materials. Older buildings were built from local materials, but improved methods of transport resulted in building materials being transported over greater distances. This created increased demand, which encouraged mechanization in the manufacture of the components of building construction.

Walls have been made from clay, concrete, glass, metal, straw, stone and timber.

BRICK

Unburnt earth

Many buildings remain that have been constructed using unburnt clay. These used the clay ground mixed with straw, either as crude bricks or as filled trenches. This type of construction is usually called 'cob' (although local variations of naming such as 'clob', 'wychert' or 'pisé de terre' do exist). Where chalk was mixed with the mud, the walls were stronger. The external faces of the building are usually covered with a lime render.

Requirements of cob
Cob walls often approaching 1m in width were constructed slowly so that each layer could dry out. The walls should not dry out completely. Do not place radiators against cob walls, or install a damp course. Water is the main enemy especially where there is high clay content, and free water should be kept out.

Clay bricks

One of the earliest brick buildings was the abbey at Coggleshall erected in 1220. The early brickmakers travelled round the country making and laying bricks to order. They searched out pockets of local earth. Many large houses have a pond which was once the source of the clay excavation within their grounds.

Some timber-framed buildings were refaced in brickwork as it became fashionable. It was an expensive and prestigious material, and was often cut or moulded into special shapes. Brick was first used for chimneystacks from the late fourteenth century. In the Industrial Revolution mechanization speeded up brick making. Prior to 1690 the clay had to be beaten by foot to break the material down and remove impurities. In 1693, a large horse-driven mincer (pugmill) was introduced to brake down the clay.

Table: Brick dimension changes

Date	Statute	Comments
1725	Size of bricks	9 × 4.5 × 2.125″
1729	Bricks made within 15 miles of London	8.75 × 4.125 × 2.5″
1839	Manufacture of Coade Stones at Coade's, Lambeth	Ornate reproduction stone decorations, usually as key stones or quoins
1770	Bricks made within 15 miles of London 8.5 × 4 × 2.5″	1770–76 bricks of this dimension only coming from London
1774	Building Act	Required no timber to be on the building face because of the fire risk
1776		All bricks made in England to measure 8.5 × 4 × 2.5″
1784	First Brick Tax	Levied per 1,000 bricks, resulting in increase in brick size
1803	Double tax imposed on bricks of greater volume than 150 cubic inches	Double tax on any brick that was bigger than 10 × 5 × 3″. After this date bricks about 9 × 4.5 × 3″ (121 cubic inches). Tax repealed in 1850. Brick dimensions remained largely unchanged until metrication in the 1980s.

Brick tiles were hung on the face of some timber framed buildings and pointed, making them look like brickwork. These tiles are found extensively in Kent and parts of Sussex, as well as in Wiltshire and Norfolk. Lewes and Brighton have many examples.

The Great Fire of 1666 resulted in timber being banned from building frontages within the City of London, through the statutes of the Tenants Law 1667 and Building Acts 1668. In the north, larger bricks still remain. The Brick Taxes resulted in timber being used on house exteriors, even though this contravened the 1774 Building Act.

Brick colour
The fashion for brick colour varied, but it was the result of the clay composition. The colour depends on the minerals present, the method of firing and the proportion of any additives. Where there is a lot of iron, the bricks are bright red (Lancashire), when burnt at very high temperatures they come out blue (Staffordshire). Brown bricks contain plenty of lime but little iron; yellow bricks contain some chalk or sulphur (Thames Valley or London Stock Brick); white bricks come from Sussex or East Anglia and contain lime but no iron; and in Oxfordshire, Berkshire and Hampshire they are more grey than white. Black bricks come from South Wales, Surrey, Sussex, and Berkshire, and are the result of the manganese in the clay.

Brick bonds

The long side of the brick is called the stretcher, and the short side the header. The various brick bonds all produce a sound wall. Flemish bond (alternate headers and stretchers in each row) was used through until the 1820s when the revival of earlier styles saw a return to English bonds (alternative rows of headers and stretchers). Rat trap bond (bricks used on their side) produces a weak but cheap wall.

Snapped headers
This occurs where the outer skin of a 'solid' wall is not fixed to the remainder of the wall. The external appearance is of a typical brick bond, but the headers do not go through to bond with the remainder of the wall. This type of wall is difficult to identify but is weaker and repairs are often required.

Brick quality

Bricks are classified in BS 6100 (Sect 5.3:1984) by their manufacturing process and suitability. The four main brick groupings are:

* Facing – face suitable for exposed conditions. The soft face of under-burnt bricks will erode. Frost action will accelerate the deterioration, causing spalling in the worst cases. Bricks with white marks (usually caused by lime blows) should be rejected for use on the exterior because of their appearance and lack of durability.
* Engineering.
* Common.
* Damp proof course bricks.

Workmanship
The horizontal and vertical deviation should not exceed 12mm per floor. Vertical alignment of joints in alternate courses is expected.

Movement
Clay brickwork expands by about 1mm/m in the first eight years, half being in the first week after being laid. A small amount of movement will be caused by reversible wetting and drying. Movement joints should be provided to allow for movement of an expansion of 10mm in every 12m of wall length. The National Housebuilders Council (NHBC) recommends 16mm wide movement joints be built in.

Sulphate attack
Brickwork is vulnerable to sulphate attack. This occurs where Portland cement, which contains trycalcium aluminate, comes into contact with soluble sulphates and water. The reaction causes expansion in the brickwork mortar joints. Fletton bricks have a high sulphate content.

Symptoms include:

• Overhanging due to expansion above the damp course, in contrast with movement in the damp wall below.
• Horizontal cracking on the inner face of the wall, due to the inner face being put in tension as a result of the expansion of the outer leaf.
• Bowing where different types of brick are used in the inner and outer faces.

Flue gases
The condensation of flue gases within an unlined chimneystack may result in sulphates being deposited into the brickwork. Sulphate expansion takes time to develop and is rarely serious within two years.

Salt attack
Soft fired bricks may fail because of spalling caused by the crystallization of sulphates behind the face. This is often repaired by applying a render face to the damaged bricks.

Sand lime or calcium silicate bricks
These bricks consist of a uniform mixture of sand or uncrushed siliceous gravel with a lesser proportion of lime mechanically pressed and chemically bonded by the action of steam under pressure. They are resistant to attack by most sulphates, but shrink on drying. The

brick surface is usually smoother than clay bricks and they are pale coloured, white, off white or cream. Their use is acceptable below the damp course; 10mm vertical movement joints should be provided above the damp proof course at intervals of 7m.

Identify the wall thickness

Measure the thickness of the main external walls. Thickness of solid brick walls will reduce throughout the height of the building. It was common for ledges to be visible within the staircase. Check the thickness of the end (flank) wall of an end of terrace property to make sure that it is more than 100mm.

Single brick walls
Solid brick walls 210mm thick are at risk of allowing water penetration where there is a gutter leak, pipe failure or an exposed location. The removal of projecting sills, copings, parapets or canopies increases the risk of water entry.

Alterations to neighbouring buildings may increase the exposure. The use of absorbent gypsum plaster (as opposed to lime plasters) can create damage to internal wall finishes.

Rainwater pipes
With solid construction, any defect or failure in the gutters or downpipes will result in water penetrating to the interior of the property. Immediate action is required.

Junctions
The junction between the wall and the roof is usually capped with a flashing. The timber will move because of moisture and thermal changes. This movement will cause some distortion at this junction.

Bows
The old rule-of-thumb maximum bulge was a third of the wall thickness. Bulges may indicate serious problems when they are found in unsupported walls, for example, where the flank wall of a house is not restrained by intermediate floors and the main staircase is alongside the wall. Small vertical movement can create a bulge ten times greater.

Book-end effect

End terrace flank walls are a point of weakness because all of the thermal expansion in the terrace can result in the flank wall distorting by up to 20mm a day. If this movement is progressive it is be called the 'book-end effect', but few failures of this type have been found.

Acoustic performance

A single brick wall plastered on both sides should reduce sound transmission by 50 decibels. Insufficient perpends mortar or using dry lining will result in inadequate noise resistance. Many party walls built with 200mm blocks provide insufficient noise reduction. The performance of separating walls will be further reduced if they support the floors. From 2001, construction standards have been raised to correct this.

Efflorescence

This is the whitish powdery surface that is often seen on the face of new brickwork. Water drying out of the brickwork will carry salts from within both the brick and the mortar. The efflorescence should stop within a year and will cause little or no damage. Bricks are classified in BS 3921 by the extent of the efflorescence that will occur: nil; slight (less than 10%); moderate (10–50%); and heavy (over 50%).

Cavity brickwork

Cavity walls started around 1920, but some rare examples date from 1848. The wall is built in two panels (or leaves) linked by wall ties. The cavity prevents water penetration from the outer face to the inner skin.

Failure is caused by:

- Poor quality construction.
- Inconsistent cavity width.
- Incorrect cavity trays.
- Failure in parapet design.
- Absence of window sills.
- Incorrect position of wall ties.
- Inadequate number of ties.
- Recessed mortar joints increases the risk of water penetrating into the cavity.
- Poor mortar fill in perpends (vertical joint).

Wall tie failure

May be caused by:

- Dirt or mortar droppings on the ties, causing dinner plate-sized patches of dampness on the inner wall face.

- Corrosion where the mortar joint cracks horizontal splitting at approximately 450mm centres. Difficult to see in the early stages.
- Colliery waste being used in the mortar, e.g. black ash, which reacts chemically with the metal of the wall tie.

For the detail of repairing methods, reference should be made to the Building Research Establishment Information Papers 12/90 and 13/90 (BRE, 1990).

Cavity trays or internal gutters

Water will enter the cavity. The amount of water increases lower down the wall. A cavity tray is needed over every wall opening to redirect the cavity water to the outside of the building usually through open perpends. Installation quality is often poor. The surveyor must check all opening for unacceptable dampness that may suggest tray failure. Defects around the opening need to be given careful thought. Repairs are expensive.

The repair liability in leasehold property needs consideration. The liability may lie with the landlord or the tenant if the wording is precise. If the tenant is being advised prior to taking on a lease the advice should deal with the following areas of risk:

- who is liable for the work;
- the risk that the lease may not identify liability;
- the cost of investigating liability;
- the cost of repair; and
- the time that it will take for the work to be done once a decision is made.

Cavity insulation
The types of cavity insulation are either rigid or fibre. The wall may be filled during construction with rigid panels, sheet material or have insulation added within the life of the property. Fibres blown into the cavity should not cause damage even in exposed locations.

Causes of failure:

- Some early methods of rigid insulation resulted in damage.
- Some built-in cavity panels caused damage due to poor workmanship.
- The improved insulation in the wall will mean that the external leaf becomes colder, creating a greater risk of water freezing in the outer brickwork.
- Frost action can cause failures to the brickwork.
- Frost damage is more likely in the open textured

blocks, which are less suitable for the external face of a building.
* Block absorption rates will vary and some movement will take place. The rate of absorption is similar to clay bricks, being between 7–12% by weight.
* One of the main factors governing water penetration is the exposure of the walls to driving rain.
* Where water is penetrating an insulated cavity wall it may be necessary to clad external wall faces or remove the insulation.
* In new construction care should be taken over the decoration of the inner face of the wall as the rigid sheet cavity fill insulation slows the drying-out process.

BLOCKWORK

The concrete block was introduced around 1850, but its use in modern construction began after 1919. Early blocks were crude and were used in the construction of internal partitions. These were brittle blocks of a grey-black and knobbly appearance. After 1955 concrete blocks began to replace bricks in the construction of the inner leaf of cavity construction and within 20 years brick manufacture dropped to half. They are cheaper to use and provide better insulation.

Blocks are manufactured with varied strengths, quality of appearance and thermal insulation.

* Insulating blocks are made with lightweight aggregate, but their load-bearing capacity is about a fifth of a dense aggregate block. (Between 6–10N/mm^2.)
* Minimum load-bearing capacity for 75mm wide block is more than 2.8 N/mm^2 (BS 6073).
* Blocks less than 75mm width are classified as non-load-bearing.
* Blocks may not be a brick dimension.
* Limitations include variations in acoustic performance, dimensional stability and water retention.

Acoustic performance

The following are indications that the wall may be insufficient.

* Plasterboard wall lining which has no insulation behind.
* An overall wall thickness of less than 250mm.
* Floors built into or resting on joist hangers in the party wall will have reduced acoustic performance.

Dimensional stability

* Movement joints of 10mm should be provided every 6m.
* The drying shrinkage of the blocks is about 1mm/m.
* Most of the shrinkage movement takes place early in the block's life.
* Thermal movement, usually reversible, causes less damage (except in gravel aggregate blocks).

MORTAR

Mortar should be weaker than the brick. If pressure is placed upon the brick, it is the mortar that should fail. Increasing the cement proportion creates harder mortar and increasing the sand proportion weakens the mortar. Where a render is applied its mix must reflect the brick type. There is a problem with inconsistent mortar mixes being used in modern construction. The actual mix can only be determined by analysis. A useful rule-of-thumb in new construction is to scratch the mortar with your fingernail. If you can gouge a groove in the mortar, you have a weak mix (weaker than ten parts sand to one of cement), if the mortar reduces your nail it is stronger. This test does not hold for lime mortars.

* Halving the mortar strength will reduce the wall's strength by 15%.
* Deeply recessed bed joints can reduce the compressive strength of the wall by up to a third and affect the lateral strength by a half.
* Unfilled perpends have a negligible effect.
* The mortar mix will affect the frost resistance of the mortar.
* Problems are more likely to be experienced with the cement rich mortar mixes.

Pointing

The pointing is the face applied to the mortar bed. This may be by the finish of the mortar bed or the application of a facing mortar to each bed. The purpose is to throw water off the joint.

* Recessed joints result in water being held in the joint and can cause water penetration.
* It is worse where the bricks are dense and impervious.

Struck joint
Formed by the mason taking the trowel and striking an angled line in the mortar joint. It is the most efficient weathered joints because water is thrown away from the joint.

Bucket handle joint
Formed by rubbing a round-ended piece of wood or
hosepipe over the joint after laying the brick to give the
joint a curved surface.

Rag joint
Formed by rubbing the face of the joint with a piece of
sacking or similar material, creating a flush joint.

Render
In certain locations, external finishes are applied to the
wall to reduce water penetration and improve
performance. For example:

- Rendering (harling in Scotland).
- Tile or slate hanging.
- Weather boarding or similar finishes.
- Painting.

Flexible render consists of a weak mix of lime, sand and
cement mortar. Without lime there is a risk of shrinkage,
which produces surface crazing. In exposed conditions
water penetration will occur. A sound render should
allow the wall to breathe and any moisture to evaporate.

When examining rendered elevations:

- Consider why the building has a rendered face (to
  improve weather resistance, to reflect the heat of the
  sun, to achieve a particular appearance or to cover
  brickwork defects).
- The adhesion can be checked by tapping the surface to
  see if there are any hollow sections. The sound is quite
  distinctive in thinner coatings, but may be more difficult
  to identify if the render is more than 50mm thick.
- Effective render faces must remain weather tight at
  door or window junctions. Shrinkage creates a gap
  that must be covered by a flashing, bead or mastic.
- An alternative is projections, mouldings, sills or a bell
  mouth projection.
- A render application to the surface of a windowsill is
  not an effective repair.
- Rendered walls should finish with a bell mouth
  projection just above the damp proof course.
- If the render continues to ground level there is a risk
  of render failure because of the damper wall below the
  damp course or capillary action allowing water to
  climb past the damp proof course.
- Pebbledash finishes are inferior and tend to require
  more maintenance.
- There is a risk of sulphate damage where applied to
  old brickwork.
- Fletton bricks provide a poor base for render.

BALCONIES

Balconies failures create a risk to their users. Balcony repair is a major item of cost in estate maintenance. Balcony construction will vary.

Stone was used in older buildings, built into the outer wall and cantilevered to form a balcony. Where the stone has a fall towards the building it increases the risk of water entry and the deterioration if any timbers support the stone.

In a concrete balcony the main risk is of:

* Corrosion in the reinforcement.
* Hand railing fixing. Different metals may accelerate corrosion.
* Inadequate reinforcement cover or water being retained in the construction increases the risk of failure. Carpet and tile on balconies absorb water and hold it against the concrete surface.

Timber balcony construction tends to have a short life, unless supported on metal brackets of adequate dimension.

STONE

Stone is classified as either igneous or sedimentary. Igneous solidified from molten, for example, granite. Sedimentary rocks are igneous rock or bone debris deposited in layers. Examples include, sandstone, limestone, and metamorphic rocks where the rock has been altered by heat or pressure.

Granite

A hard, durable and strong rock but difficult to work and expensive. The quality of joint workmanship is important.

Sandstone

Strength can vary between weak friable and porous, to strong and durable. It is prone to having soft pockets because of the lack of uniformity. Erosion may not be uniform. It is chemically inert and has withstood pollution. Its colours can vary from grey, through pink to the reds of Devon, Liverpool and Glasgow.

Limestone

Classified in six groups from A to F, with A being the most durable and suitable for use in most exposed locations. It has been suggested (BRE, 1988) that pore size is relevant to performance; the larger pored stones, being less able to hold water, perform better. Weather resistance varies, with Purbeck being more resistant. Some limestone is porous and vulnerable to water penetration, e.g. Bath. Limestone's colour will vary from white to the golden walling of Bath and the Cotswolds.

Durability of stone

Stone is strong in compression and weak in tension. For that reason most stone lintels bridge short spans.

Causes of decay

Even though pure water is harmless to stone, water is the main cause of damage. A dry stone is less likely to decay, but stone cannot be completely waterproofed without changing its appearance. The application of silicone treatments has caused stone damage.

Causes of failure:

- Incorrect bedding.
- Weathering.
- Rainwater.
- Wind erosion.
- Pointing failures.
- Erosion.
- Rising dampness.
- Salt entry.
- Soot encrustation.
- Birds.
- Seeds.
- Ivy.
- Creeper.
- Blisters.
- Design of mouldings.
- Efflorescence.
- Frost action.
- Corrodible metal clamps and fastening.
- Association of dissimilar stones.

CONCRETE

Concrete is a mixture of cement, water, fine and coarse aggregates, and air. Cement and water form the glue that coats all fine and coarse aggregate. The concrete will continue to gain strength after it sets.

The main classification of concrete is by weight. Lightweight aggregates create a lightweight concrete, which can also be produced by aerating the concrete. Concretes above 200kg/m^3 are classified as dense concrete. Concrete without fine aggregate is called no-fines.

Requirements of the mix

* Quality.
* Batching.
* Water.
* Aggregate.
* Temperature and moisture control.

Concrete is strengthened by the addition of steel bars or mesh to resist tension. The reinforcement may be either pre- or post-tensioned. The reinforcement is either stretched before or after being cast into the concrete and the tension then released. This increases the spans achieved by reinforced concrete. Concerns are now being expressed about the durability of the reinforcement steel. Where high tensile steel has been mixed with nickel, nickel sulphide has eroded the strength.

Concrete usually has to meet one or more of four functional requirements:

(1) strength;
(2) durability;
(3) fire protection; and
(4) thermal insulation.

Its properties depend on both the correct selection of materials and good site practice.

* Most concretes are made with ordinary or rapid hardening Portland cement.
* Ordinary cement develops its working strength after 7–14 days.
* It continues to grow in strength and the rate of hardening is accelerated or retarded by the temperature.
* It has a low resistance to attack by acids and sulphates.
* Rapid hardening cements develop their strength much more rapidly although they do not set faster.
* The sulphate resisting cements are used where there is a risk of sulphate attack in certain ground conditions.
* Extra rapid hardening cements are not recommended for use in reinforced or pre-stressed concrete. The chloride content increases the risk of corrosion. Because of the speed of the accelerated hardening, the concrete has to be placed and compacted within 30 minutes of mixing.

- Ultra high early strength cements contain a higher proportion of gypsum and develops early initial strength. There is little increase in strength after 28 days. They are suitable for use in reinforced or pre-stressed concrete, but they suffer more creep.
- Higher-heat Portland cements tend to be prone to surface crazing or fractures in the cement after it has gone off. Lower heat cements are used where large masses of concrete have to be laid and it is important that they develop their strength more slowly to avoid surface cracking.

High alumina cement (HAC) develops very high early strength, which enables shuttering to be struck within 24 hours of the initial pouring. If the temperature of the concrete exceeds 30°C, it will probably lose a substantial proportion of its eventual strength. HAC has a higher resistance to attack by acids and sulphates but the water to cement ratio needs to be carefully controlled because it influences the eventual strength of the concrete.

HAC failures

- In July 1974 the Department of Environment stated that buildings over five stories incorporating HAC must be regarded as suspect.
- The greatest risk lies in the use of pre-cast and pre-stressed isolated roof beams.
- In warm and moist locations chemical action may occur. This may cause further strength reduction.
- The risk of structural failures in spans up to 5m is small.
- HAC continues to increase strength for some months after casting but then its strength reduces.
- Strength reduction is greatest when there was a high water content in the mix and high temperatures occurred during curing.
- Minimum strengths occur between five and ten years after mixing. It is reasonable to expect HAC to maintain its strength after the period of ten years, and some improvement may occur.

An external examination for defects in HAC may reveal signs of failure such as:

- Excessive deflection.
- Lateral bowing.
- Cracking:
  ○ The cracks may take the form of sheer fractures or flexing cracks. Special attention should be paid to any areas that have been subjected to high moisture penetration, e.g., roof leaks or excessive condensation.

- HAC is vulnerable to chemical attack from plasters and woodwool slabs.
- Both white and black markings may be indicative of chemical attack.
- Samples are required for a chemical examination.

Most concrete failures result from the following:

- Reinforcement failure due to inadequate cover.
- Carbonation. The alkali nature of the concrete becomes acid through contact with weak acid moisture.
- Advanced corrosion having already started prior to the installation of the reinforcement.
- Chemicals in the concrete mix reacting with the reinforcement.
- Failure to select the correct type of concrete for the purpose required.
- Failure in permanent shuttering, or its reaction with HAC.
- Impurities in the materials used.
- Excess water used in the mix.

Absence of provision for movement

Failures will occur in concrete construction if expansion joints have not been installed within the original framework. The following points must be taken into consideration:

- Design of the movement joint must relate to the maximum anticipated movement.
- Frequency must relate to both horizontal and vertical movement.
- Type may rely upon sealants or baffles.
- Integrity of the structure and the weather resistance of the envelope must be unaffected.

Carbonation

Concrete is alkaline and protects the steel, inhibiting rusting. Carbon dioxide and industrial pollutants in the air can mix with rainwater and form carbonic acid, which will neutralize the alkalinity and increase the risk of corrosion beginning.

Uncarbonated concrete has a minimum pH value of 12.6. Carbonation will reduce that to about 8.0pH. A neutral state is around 6.0pH. The progress of the carbonation starts from the outside and moves inward at a reducing rate. With some cement to water ratios it can achieve a depth of 50mm in 15 years. This may affect the reinforcing rods, which often have a cover of concrete of between 50–75mm.

Carbonation will be more rapid in:

* Porous concretes.
* Concretes with low cement content.
* Concrete with a high water to cement ratio. Concrete with water to cement ratio of 0.80 will be twice that with 0.60.

The permeability of the concrete will depend upon the:

* aggregate which was used in the original mix;
* cement content; and
* water to cement ratio.

Use an alkali-sensitive stain based on phenolphthalein to test the depth of carbonation. A small piece of concrete is broken off (not cut with a saw) and sprayed – if not fully carbonated the liquid will develop a purple red stain. This is unreliable in HAC.

Alkali-Silica-Reaction (ASR)

Caused when alkaline chemicals present in cement react with silica. A jelly-like substance forms around the stones, absorbing water and expanding. This usually causes cracks in the concrete. If water gets in and freezes, the cracks become wider and let in still more water. The problem can be avoided in new building work by using low alkali cement or by avoiding the use of chemically reactive aggregates.

Ground salts

Sulphate-bearing soils can cause damage to concrete if there is inadequate cover to the reinforcement.

Physical damage
The most common causes of physical damage are:

* Frost action.
* Impact damage.
* Stress.
* Fire damage: this damage should be obvious and the agencies of damage evident.

Stalactites
Calcium carbonate comes from water leaching out the hydrated lime from Portland cement which then reacts with the carbon dioxide in the air. Worry about where the water is coming from and correct the leak. The liquid is caustic.

Steel corrosion: additives in the mix

Salts containing chloride ions were added to the concrete to speed up the setting. They destroy the protective surfaces of reinforcement. The corrosion occurs as the result of micro galvanic cells. If chloride of iron (such as calcium chloride) was added, the risk of corrosion increases.

Exposure to caustic agencies

Chloride attack may result from salts being added to the concrete. This may have occurred because sea water or unwashed sand and aggregate from marine sources were used in the mix. The adding of calcium chloride in excess of 2% by weight of cement will increase the risk of corrosion.

Shrinkage

The addition of calcium chloride leads to an increase in drying shrinkage.

Woodwool slabs

Problems have been found where woodwool was used as permanent formwork due to:

* it being easily crushed;
* severe honeycombing because of the damping effect during concrete vibro-compaction; and
* its reaction with HAC.

The cost of the remedy

Consider the economic implications of the need to repair. Consider the cost and extent of the work, and the value of the building before and after the repair. Is it worth repairing?

Repairs

* Carbonation treatment. Anti-carbonation coatings can be applied but have a life expectancy of about ten years.
* Epoxy resin mortars. Don't provide alkaline protection or an impermeable barrier. Steel reinforcement must be first treated with a protective primer coat. There is low shrinkage and rapid hardening.
* Re-alkalization. The reinforcement bars are connected. The concrete is exposed to an alkaline solution, usually sodium carbonate, suspended in water on a mat. Low voltage current is passed through

the reinforcement so that it becomes an anode, thus attracting the alkaline particles that are drawn into the concrete from the solution in the mat. Limited life expectancy of about ten years. A wet and messy process that takes four to seven days for each area of treatment.

* Small repairs: cement modified with polymer latex. The exposed and cleaned reinforcement is treated with a highly alkaline passivating system. The concrete is replaced as a dry mix, being a putty consistency.

EXAMINING CONCRETE SYSTEM BUILDINGS

System building was introduced into the UK in the 1960s to meet the demand for the rapid development of a large number of homes. Most are controlled by RSL (Registered Social Landlords) or have passed into the private sector through the Right-to-Buy policy of the 1980s UK government.

The systems involved the fixing together of pre-cast concrete components.

Causes of failure:

* Systems joints were too complicated.
* Poor workmanship.
* Joint construction was poor.
* Lack of management of the construction process.
* Use of additives in the concrete.
* Blight on re-sales.
* Poor insulation.

INSPECTING CONCRETE CONSTRUCTION

Identification

The initial identification will be based upon panel size. Most construction types are divided into small or large panel systems.

Research

One of the best sources of information is the series of booklets produced by the Building Research Establishment on condition of most forms. They include Airey, Boot, Cornish, Dorran, Myton, Newland, Tarran, Orlit Parkinson, Reema, Stent, Unity, Winget, Underdown, Wates and Woolaway.

Common problems

Most are vulnerable in two places – the concrete and the reinforcement. Examine concrete for both unsuitable additives and carbonation. Calcium chloride was frequently used. Some concrete was aerated or breeze concrete. This is less resistant to the weather and a greater cover to any reinforcement is required.

Tests of the concrete structure:

* Cover meter test.
* Vibration testing – to locate the voids in the concrete.
* Schmidt hammer – limited guide to the quality of the concrete.
* Ultrasonic test – concrete strength indication.

Repairs

Acceptable methods of repair have been designed for most systems. The repair cost varies for each type.

TYPICAL LIGHTWEIGHT CONCRETE CONSTRUCTION

Schindler Hawksley

Brick faced timber frame with integral concrete site poured frame. Check for metal stay under stairs. High repair cost for replacing all external walls.

Airey

Two-storey with concrete ship-lapped exterior.

Failure: the concealed columns have a central hollow pole reinforcement with minimal concrete cover. Repair involves replacement of the structural frame spine, party and external walls.

Boot

Render face with concealed column and infill panels.

Failure: use of dense aggregate concrete. Carbonation is minimal in Scotland, but up to 25mm in England.

Orlit

These may be rendered or with a column and panel exterior. Block panels with straight joints with flat or pitched roofs.

Failure: chloride concrete and advanced carbonation.

Cornish

Visible columns with infill panels. Column spacing about 1m.

Failure: high chloride content found in south-west England. Carbonation has reached most steel reinforcement.

Unity

One and two storey with block panels with straight joints. Some are render finished.

Failure: Chloride used in concrete, steel corrosion common.

Wates

One, two and three storey buildings, with storey height panels of varying width and ring beams to each floor. Corners either butt or mitre joints.

Failure: Low chloride content. No visible signs of deterioration.

Sealant Failures

Causes of failure
Approximately 80% of sealant failures occur for one of the following reasons:

* sealant applied too thinly;
* not having been laid on a properly primed surface; or
* the incorrect sealant used.

Application

* Depth should be twice the width of the joint.
* Sealant must adhere only to the intended surface.
* Sealant flexibility must cover the range of joint movement.
* Triangular fillets' diagonal face not less than 10mm and surface convex.
* Gaps over 5mm need back-up material.
* Sealants may react with certain materials.
* Incompatible materials isolated with bond breaking tape.
* Some sealants require primers applied before application of final sealant coat.

Types of sealant

Sealants in common use are oil-based or oleo resinous:

* butyl;
* acrylic;
* polyurethane;
* polysulphide;
* silicone; and
* bitumen.

Failures

There are two types of visible failure with sealants. These failures are:

* adhesive; and
* cohesive.

Adhesive failures
Sealant loses adhesion. Curved sealant bead pulls away from the joint interface due to:

* Poor joint preparation.
* Dust in joint interfaces.
* Incorrect primer used.
* No primer used.
* Sealant applied to damp surfaces.
* No joint backing.
* Lack of tooling during application causing incorrect sealant configuration.
* Comprehensive thermal movement prior to sealant curing.
* Joint movement in excess of sealant capacity.
* Sealant depth inadequate.
* Sealant incompatible.

Cohesive failures
Occurs when the material splits because:

* Thermal movement exceeds joint design.
* No joint backing present.
* Inadequate central depth.
* High thermal movement occurring prior to full cure.
* Joint design unable to cope with extent of movement.
* Poor quality sealant.
* Inadequate bond breakers.
* Entrapment of air behind the sealant.
* Poor mix of component producing unstable sealant.

Repair work

Remove all material, clean up the joint and replace the seal. If the sealant was oil-based removal is essential.

Over-capping of butyl has failed because the butyl migrates from the joint and contaminates adjacent areas. The specification of over-capping is the work of specialists because of this risk of rejection.

Table: Sealant diagnosis

Observation	Defect
Sealant cracked and dry	End of life. How easy will it be to clean off the sealant before reinstatement?
Sealant broken away from face of one or both sides of the joint	Sealant either deteriorated or unable to deal with the movement within the joint. Possible poor preparation. Design replacement to handle predicted movement
Sealant compressed and pushed out of the joint – possibly linked with crazing to the edges of the panels	Joint too small and inadequate provision made for movement taking place. A detailed examination behind the panels will be required
Sealant shows signs of transverse movement, the surface having ripples	Movement has taken place along the line of the joint. This is not likely to have been predicted in the design and suggests a failure in the panel or frame fixing. Detailed examination required
Sealant not in contact with the panel infill	The panel has been incorrectly fixed. In glazing this may be the result of spacers not having been used – this may have caused the failure of the double-glazed panel
No visible defect or distress	Is the joint under-deployed? Is sealant placed on the face of an imitation joint

Metals

Aluminium – durable with high strength to weight ratio. Co-efficient of linear expansion is twice that of steel. It develops a whitish oxidized surface. It is virtually non-magnetic. It should not come into contact with lead. Electrolytic reactions will adversely affect aluminium, in damp conditions, if in contact with copper or stainless steels.

Copper – used in roofing, electrical installations and wood preservative. Water washing off copper will leave a green stain. Copper in suspension in water will damage galvanizing on pipework. Manganese salts will pit and erode the copper. Some bath salts contain extracts, and copper bath traps are particularly vulnerable.

Lead – highest level of thermal movement of all the metals, and its fatigue resistance is relatively low. The greatest weakness is at the junction of vertical and horizontal surfaces and on the curve of a lead roll. It is vulnerable where is comes into contact with lichens, peat and western red cedar, and if it remains wet and comes into contact with Portland cement or concrete.

Cast and wrought iron – strong in compression and weak in tension. This makes it good for columns and weak for beams.

Wrought iron – moderately strong in tension and easy to work. It replaced cast iron from the mid-1850s. It is tough and resistant to impact. Repairs are usually undertaken by bolting or electric arc welding.

Steel – alloy of iron, carbon and manganese. The amount of carbon in the steel increases the hardness and the brittleness. Brittle fractures occur with the introduction of all weld-jointed buildings from the 1940s.

The weld – success influenced by climate and steel's reaction to being melted. The steel must be dry at the time of the weld. There is a risk that the weld may make the steel more brittle. The joint may be a fillet or a butt. The fillet results in the weld being added to, whilst the butt joint involves the molten metal being poured into a groove at the junction of the two plates, so that the resultant finish projects above the surface by a slight amount. The quality can only be tested by using sonic equipment or by X-ray equipment.

Stainless steel – developed in 1913, an alloy of steel and chromium. It has a high resistance to corrosion. It is vulnerable to acid damage, and may in large quantities accelerate damage to mild steel. It is stable, its thermal movement being low and it is non-magnetic.

Zinc – used for coating other metals, or cheap roof covering. The metal has too short a life to be left unprotected. The metal is vulnerable to chlorides, sulphates, soluble salts and gypsum plasters. Contact with damp oak and western red cedar and copper must be avoided.

Corrosion of metals

Corrosion usually results from a complex electrical action where dissimilar metals come into contact.

The further apart in this list the greater the risk of their being in contact.

Table: Table of metals and their interaction

Anodic Magnesium
Zinc, including galvanized coatings
Aluminium
Cadmium
Copper–aluminium alloys
Iron and mild steel
Chromium
Lead
Tin
Nickel
Brass
Bronze
Copper
Stainless steel
Cathodic silver

Windows

Thermal performance

Windows in new buildings must, as part of the performance of the building enclosure, achieve a target U value. Replacement windows should not have a U value less than 2. To achieve that the overall double-glazed panel will be 24mm (16mm cavity) and low emissive glass will be used. The thermal coating with glazed panels is often affected and goes cloudy in direct sunlight but changes back to clear afterwards.

Sheet glass was first introduced in 1773 but from 1840 it became commercially available in larger panels. The sliding sash window was developed from the 1700s and continued in use until the twentieth century. By 1913 continuous production of sheet glass was possible. By 1923 polished glass was being made; and in 1959 Pilkington invented float glass.

The window head

The building above the window opening will be supported by a lintel, which can be made from stone, timber, steel or be formed as an arch. The materials used will determine the width of the window opening. In brickwork a cut brick arch will easily span up to 1.5m. A flat stone lintel may be limited to little more than 1m width, a precast concrete lintel to around 2m, but timber lintels will span up to 4m, and steel over 10m. Look at neighbouring properties for any indication of construction weaknesses.

Brick arch

Some brick arches are backed by a timber lintel. Slippage is common and may be due to:

* erosion of soft brick cut to form the arch;
* movement removing the arch's support;
* widening the opening; or
* loss of pointing or the key stone.

Bressummer

A timber beam over a substantial opening on the front of a building. The timber often bends under load or deteriorates. It is vulnerable to water penetration. Look for movement in the brick courses.

Lintols

The use of iron and steel started from the 1870s, but the use of metal lintels did not become standard until after WWII. Steel or iron channel sections were set into each side of the window, and the brickwork was continued within the web. Look carefully for the sign of splitting in the mortar bed to the window head, this may suggests corrosion.

The windowsill

Stone sills of pre-twentieth century property were made from soft stone which weathers poorly. Movement within a building will cause fractures, the stone not being capable of taking stresses set up by the redistribution of weight. The windowsill should have a drip to the underside. Check the following carefully for failure:

* the skirting below the window;
* the wall below any window;
* the window board; and
* the frame for any sign of deterioration.

Sash window

The vertically hung sliding sash window is two counterbalanced frames which slide within runners set in the sash box. Changes in glass thickness imbalances the sash weights. Sash cords were waxed rope or chain, but nowadays wire reinforced or nylon reinforced rope are used.

Casement

The casement window, introduced in Tudor buildings, remains in use. The hinge of metal framed casements are prone to failure. The diamond-shaped leaded lights enabled some movement to take place within the frame without fracturing the glass, but bulges had to be supported by metal bars across the window.

Metal window frames

The modern metal windows are also prone to hinge failure. Minor movement of the property often results in distortion of the main frame. Rusting of the metal casement can result in the glass fracturing.

Brickwork

* Brick jambs wrap around the front face of the timber sash box so that much of the window frame is concealed.
* The render fillet to the brick jamb will fractre if movement occurs.
* Vertical damp course is inserted at the junction of the outer and inner leaf.

The window should be recessed by at least 60mm to cover this junction.

Ventilation

The glass area must not be less than 10% of the floor area and the area of ventilation should be not less than 5% of the floor area. A door does not count as a window and therefore does not form part of the ventilation calculation.

GLASS

Glass is made from a mixture of sand, soda ash, limestone and alumina. It is vulnerable to damage by acids and alkalis, for example, caustic soda or water running off concrete. Thermal movement is around 0.3mm/m for traditional glass, but may be over 1mm/m for solar control glass.

Glass should be able to deflect up to 1/125 of its span. Patterned and wired glass is much weaker, although when broken much of the glass remains in place, as the thin metal wire holds the shards. It is vulnerable to impact damage. Opaque glass is weaker than clear glass, and solar control glass is liable to thermal stress and cracking.

Toughened or tempered glass is more flexible than annealed glass and is over four times stronger. It suffers from optical distortion, and may be able to be identified because of the roller waves that may be seen on the surface reflection. Nickel sulphide may cause damage to 1% of toughened glass if:

* the nickel sulphide is unstable;
* it is larger than 0.03mm; or
* it is located in the central tensile zone of the glass (heat-strengthened glass is not vulnerable to nickel sulphide failures).

The published failure rate during heat soak testing is around 0.6%. Failures occur after the first year and the risk of failure is then ongoing for over 20 years.

In critical locations (for example, doors, side panels to doors or partitions, walls, or windows) glass must either: break safely; be in small panes of ordinary glass; be thicker than ordinary glass; or be protected by a permanent robust screen.

Safety glass

Safety glass includes toughened glass categorized as Class A, laminated glass available in Class A, B or C; or wired glass (also called Pyroshield safetyclear/ textured) categorized as Class C.

'British Standard 6206: 1981 (1994)' requires that each piece of safety glazing used within 'critical locations' should be marked with:

- the British Standard number 'BS 6206';
- the type of glass used, i.e. 'L' for laminated, 'P' for plastics, 'T' for tempered (toughened), 'W' for wired or 'SFB' for safety film backed;
- the category of safety glass used; and
- an identifiable name, trademark or other identification mark of the manufacturer.

Domes and rooflights

Failure is common at the roof junction with the kerb of the rooflight. Metal framing is prone to the same deterioration discussed for metal window frames, the glazing being cracked by the corrosion of the metal. Where ventilation is provided by opening and closing panels, age reduces effectiveness.

Dormer windows

Many leak and are difficult to examine. The junction of the dormer and the roof surface may not be accessible, yet is a vulnerable joint. Check the integrity of the roof structure.

Sliding doors

Patio doors slide within grooves set to the top and the bottom of the frame. Small wheels are set in the door panel. Check that window locks prevent the doors being lifted from their frame.

The bay

The bay projection often moves independent of the main building. Before considering if the bay is moving, look at the construction. If the bay is faced in stone, it is common for the framework of the bay to be timber with the stone inadequately secured.

Louvers

These slatted windows are a security risk. The slats are only clipped and easily slide out. The use of such windows should be backed with adequate security grills.

Access

Check that first floor windows can provide access for furniture if the staircase is tight or winds.

Cleaning

Check that there is safe access for cleaning. In domestic property damage is often caused as the occupant stands on any projection on the outside of the building. If a roof surface will not carry the weight of a person warn of the risk.

CURTAIN WALLING

The most common type of external non-loadbearing envelope for a building is the curtain wall. It consists of an exterior skin backed with insulation, a vapour barrier, sound-deadening materials and an interior skin. This external skin is either fixed to a frame or each panel is fixed to the structure of the building. The panels of this exterior skin may be made of glass, metal (stainless steel, aluminium, bronze), masonry (brick, tile), concrete (pre-cast concrete panels, concrete), or stone (limestone, marble, granite).

Joints

The joint design either keeps water out or collects water and discharges it.

Front sealed
In a similar way to the domestic window frame, the panels are sealed at their junction with their frame with a two-part polysulphide or silicone sealant. Site applied sealants are variable in the quality of their application, and the system relies upon a good quality of workmanship. Because of the risk of on site application failures, many designs rely upon a neoprene or silicone gasket.

Drained and ventilated
A series of recesses collect water and baffles redirect water at the panel joint. The efficiency relies upon the durability of the materials used, including the quality of their fixing. A baffle, suspended within the joint, is vibrated by wind movement which may damage the fixings.

Pressure equalized

Pressure differentials can allow moisture to be sucked into the framework. By balancing the pressure behind the external skin is avoided. The system relies upon effective seals and controls to avoid the expansion of trapped air creating future suction when a rapid temperature change takes place.

Failures in the envelope

* Sealant failure.
* Air leakage round glass or panels.

Fire risks of cladding and curtain walling

Insulated cladding is at risk because it can help the spread of fire. A fire in a block of flats in Irvine, Ayrshire, Scotland in June 1999, spread through the void to the rear of the external cladding. The report's recommendations will require new standards for cladding, and cladding to buildings of ten storeys must be made safe.

Systems of external cladding mentioned in the DETR report included those based on render applied to insulation, rain screen cladding where the panels are held off the face of the building, and pre-formed infill systems where the cladding is delivered with the insulation attached.

Floors

The survey must include the identification of risk for the defects that may be present within the areas that are concealed. The examination of the floor should start, after consideration of the possible construction, with a walk through. Before the floor is evaluated it will help if the functions of the construction are considered.

MAIN FUNCTION: CARRY THE LOADS IMPOSED

The floor will have to have an adequate deck that can transfer the loads to the beams that transfer the loads to the beam supports. Those loads may be static or be dynamic. The floors may have to deal with weight, vibration, lateral thrust, impact, and changes brought about by humidity or water condensing. The services placed in the floor may produce temperature variations of both heat and cold. The drying shrinkage does cause problems if the design did not provide for the reduction in timber dimension.

Secondary functions

+ Keeping out dampness.
+ Providing a suitable surface.
+ Providing thermal insulation.

FLOOR CONSTRUCTION

The construction of floors may differ between the ground floor and upper floors within the building. The floor may be constructed of steel, timber or concrete.

The dimension of the timber joists is important. Undersizing is a common problem with new construction. With the growth in the use of particleboard floor sheets, the surveyor is hampered in the inspection. The table below sets out the requirements for the installation and the use of such materials.

Table: Common timber sizes for spans

Timber size (mm)	Joist centres (400mm)	Joist centres (450mm)	Joist centres (600mm)
50 × 100	2.00m	1.90m	1.54m
50 × 150	3.13m	3.01m	2.66m
50 × 200	4.15m	3.99m	3.48m
50 × 225	4.65m	4.45m	3.88m
38 × 100	1.80m	1.64m	1.28m
38 × 150	2.91m	2.76m	2.40m
38 × 200	3.82m	3.61m	3.13m
38 × 225	4.25m	4.02m	3.50m

Loading based on class SC3 timber at a load not exceeding 0.5kN/m²

If a steel joist has been inserted, the bearing of the steel will need a padstone, the bricks will be insufficient to support the joist end.

Where timber joists bear on an outside wall they may deteriorate, particularly in solid brick walling. Timber joists may have been supported on metal hangers, which are hooked on the inner leaf of the wall. These joist-hangers do deteriorate over 40 years or so, may be inadequately installed, and may damage the wall if bending takes place in the joists. Floor joists should not be bedded in the party wall, as there is a risk that the timbers will carry noise from one building to the other.

In pre-1900s houses, floor joists were placed on horizontal wall plates that ran continuously through the walls. These spreader timbers were intended to redistribute the load evenly through the wall. They are prone to decay.

The examination of the ground floor construction should be made wherever possible. If there is a cellar, access may be provided to the sub-floor. In many houses in the south-west and north-east of the UK, access hatches are provided in the sub-ground floor in the under-stairs cupboard.

Fire resistance

In multi-occupation buildings it is essential that an adequate firebreak is created by the floor construction. Modern buildings with a concrete floor will achieve this.

Whatever the construction, the surveyor must know what would achieve a requisite standard of fire resistance.

Table: Standards of fire resistance

One hour separation	Ceiling	Floor
Timber floor joists at least 50mm wide	30mm plasterboard	Minimum T&G 15mm
Concrete not less than 95mm thick	At least 20mm cover to reinforcement	Any
Half-hour separation		
Timber floor joist at least 37mm wide	12.5mm plasterboard with taped joints or 9.5mm plasterboard with 10mm lightweight	Minimum 15mm plaster finish plain edged boarding with 3.2mm standard hardboard nailed to floor boards

Ground floor: concrete

The concrete ground floor, installed by pouring concrete into the permanent shuttering of the external walling, was seen as a cheap and economical solution by developers and contractors. Defects are common in this type of construction. Because of the problems there has been a move towards the use of concrete suspended flooring in new development.

The main defects in solid floor construction result from:

* inadequate compaction of the hardcore;
* inadequate linkage of damp courses within the floor and the walls;
* inadequate allowance for clay heave and where insulation has been installed; and
* inadequate thickness of cement screed being applied.

The main disadvantage of constructing solid flooring is the high cost of repair where a defect needs remedy. Most such flooring does not have insulation added at the time of construction. The result is that the floor surface is cold, leading to the formation of condensation.

The installation of services is more complex. The floor design will have to provide for the arrangement of pipework.

- Alterations to encased services are expensive.
- Errors in layout cause extra cost.
- Layout details may not be available when casting the floor.
- Workmanship errors may add cost.

Any upward or downward movement of the ground will directly affect the floor. The downward movement of the floor slab may occur because of the compaction of the hardcore or drying out of a clay soil. This is more likely to take place in the first few years after construction.

Hardcore

The purpose of the hardcore is to ensure that the ground is loaded evenly. With care a clean dry and level working surface can be created prior to the laying of the concrete. Hardcore may include bricks and clay tile, clinker, gravel, quarry waste and shale. It must not include any material that may swell, such as plaster, pieces of wood, colliery shale or thermal blocks.

Sulphate attack does happen where the hardcore has contained water-soluble sulphates. These are leached into the concrete floor slab, which then expands. The floor domes and a distinct hollow sound can be heard when it is tapped. Where the cause is colliery shale, other buildings will be affected in the area.

The hardcore is also capable of expansion. The use of plaster which may contain unhydrated lime, materials with a high clay content, or waste products that will react with oxygen and water to expand, have all contributed to the deterioration of concrete slabs.

Suspended floors

The suspended floor is easier to repair, although that is a very negative way of deciding which method of construction one should adopt.

- Because the components are above the ground they are unlikely to be damp.
- They do not require the installation of a complex damp course.
- They allow the ground below to breath provided the sub-floor is ventilated.
- They are more flexible when it comes to the installation of services.
- They are more flexible for use on sloping sites.
- The concrete beams do have a camber. Floors of particleboards on insulation, rock and creak.

It is essential that the floor void below the floor be ventilated. If the concrete beams become wet there is a risk of early failure.

Intermediate floors
The intermediate floors in a building may be formed as a timber deck or as a concrete beam and fill system. Noise spread is greater with timber or frame construction than with mass concrete.

The timber floor deck is vulnerable to being damaged by services installers, who may cut excessive notches through the support joists to pass pipes and cables. Whilst notches close to the bearing are survivable, notches in the mid-span of undersized joists may cause failure.

The floor should work as a platform, where each joist is braced against the next. Solid blocking is the usual way to achieve this stiffening of the platform, although the boards add strength to the floor. With no blocking in place, distortion of the timber joists will take place.

Materials

Some of the more common decking materials have caused failures where they have been installed incorrectly.

Particleboard floor decking
The use of particleboard in domestic flooring has increased in recent years, both as an alternative to timber boards in joisted floors and as the deck over insulation. Moisture retention in the flooring will contribute to deficiencies. Even some of the 'moisture resistant' grades are unable to withstand prolonged wetting.

Chipboard is the most commonly used particleboard for domestic flooring, however, oriented strand board and cement bonded particleboard is also suitable for domestic application.

The most common failures result from water entering the boarding and the boarding expanding. The boards will also shrink if they were damp when laid. Prolonged wetting will lead to a permanent loss of strength. Inadequate fixing can result in cupping, or lifting at the board edges due to differential moisture movement within the thickness of the boards. Boards applied to floating floors are particularly vulnerable.

Arching of wood block floors
Moisture penetration into a woodblock floor will cause expansion of the timber. An expansion strip should be provided, but will not cope with extensive dampening of the floor blocks. Some floor cleaners and polishes fill the pores in the cork and reduce their ability to take up expansion in the wood block.

Timber floor finishes – hardwood panels

The appearance of natural wood flooring is popular. Some of these products are not wood at all, but paper printed with a wood pattern fixed to a thin particleboard and topped with a thin surface. As with wood blocks provision is needed for expansion and contraction to take place. Examine the skirting junction to see how this junction has been formed and ensure that an expansion gap has been provided. Check that there is sufficient space around heating pipes passing through the floor surface for the movement not to cause pressure on pipe joints.

Timber floor finishes – sanded boards

If the floor has been sanded, make a careful check to see if any damage was caused. In particular, check service pipe connections to radiators for the amount of metal removed in the process, or any cutting into electrical conduit. The floor will be noisy, and may be in contravention of a lease in apartments or flats.

Timber flooring: general

The moisture content of the surface of the floorboards must be tested. Bounce lightly on the balls of your feet whilst examining the floor, particularly at the junction with external walls. Feel for any semblance of a weakness in the support timber where they are bedded in the outer wall. Check carefully for any gap between skirting and floor surface. The failure of the support timbers, or of a timber plate will cause the floor platform to drop. This expands the gap between floor and skirting.

Cement screeds

The quality of the base upon which the screed is applied will affect the screed. Sometimes screed is laid over timber flooring in kitchens to form a base for a tiled floor. This will increase the floor height. As a result the weight on the floor, or thermal movement in the timbers, movement may be transferred into the screed and cause cracking of the floor finish.

Most screed failure will be related to the preparation of the concrete base. The concrete slab should be wet when the screed is applied, unless a damp membrane is to separate the two surfaces. The screed should be given at least a week to cure or there is a risk of cracking taking place. Because it is one of the last elements to be installed, its curing (and drying) is too often accelerated. Where the screed has dried too quickly, there is a tendency for curl to occur around daywork joints.

The mix of the screed may lead to total or isolated failures. If inadequate mixing of the materials has taken place there may be isolated sandy pockets. These can be cut out and repaired, but their presence could be the indication of a more widespread failure to the flooring. Tap the floor to hear if there are isolated hollow patches that may suggest failures.

Embedded services will contribute to failures unless care has been taken in designing their installation. The warmth, expansion movement, or lack of sleeve provision to heating pipes within the screed will cause a line of cracking immediately above the pipes.

Electrical under-floor heating tends to be expensive to run. Whilst it is not used nowadays, there are still some buildings where it has been installed. If the property is leasehold, the lessee may have a liability to reinstate the electrical heating if it fails during the currency of the lease.

Tile surfaces

Provision must be made for quarry tiles expansion and contraction upon laying. Thermal movement where tiles are exposed to direct sunlight can cause the bond with the substrate to break. Test the floor for dampness and see what provisions have been made for expansion.

PVC tile shrinkage may be caused by the loss of plasticizer in the tiles. This can be caused by incorrect adhesive, which if the cause, will remain sticky.

Quarry or other ceramic tiling on balcony or patio roofs may result in noise transmission to any accommodation below. Impact noise will be considerable. If they were to have been added to the construction in a leasehold property, a lessee may be liable for their removal and the application of a sound deadening material.

Access floors

Office accommodation may be provided with an access floor. Where the floor is laid on a cambered concrete floor, note that the clear space to the middle of the floor will have the minimum height. Measure here to ensure space is adequate.

Check to see if the supports for the tiles, usually in height adjustable props, which sit on the concrete screed, have caused any damage. The failure of a screed at this point is expensive to remedy.

Where such a raised floor has been fitted to an existing building check that the junction of the floor to outer levels, such as staircases and corridors, is evenly graded and that there are no steps. Make sure also that the integrity of the fire standard of the doors is not impaired by any ramps.

Alterations

Errors in alterations often result in revised load paths within the structure. The creation of rooms within roofs is often done badly. Even if the joists are big enough, make sure they are supported on the walls and not on the original ceiling joists. The space created may not rate as a room, because of poor headroom, insufficient floor space, lack of ventilation, the size of windows, the absence of consent, or the failure to provide adequate means of escape.

Security
The floor in multi-occupational buildings is also a security barrier. Consideration should be given to the effectiveness of the floor as a barrier to entry.

Sound
Noise spread from one flat may be at unacceptable levels because of the floor construction. The floor should be considered for the likely risk of it being an insufficient barrier. Most flat leases require floor surfaces to be carpeted to reduce impact noise.

Staircases

If movement has taken place the evidence is usually present around the staircase.

- Check the junction between a staircase and a flank outside wall.
- Check the junction of the staircase and each landing.
- Make sure that the fixing of the handrail is safe and secure.
- If there is a skylight above the staircase run your fingers over the rail below it to see if there has been any water entry.
- Check that the balusters and posts are sound.
- Make sure there is insufficient gap for a small child to fall through.
- Make sure there is an even going.
- Consider the width of the staircase and whether it can be used to carry large furniture or bedding to the upper floors.
- Examine the underside for signs of beetle infestation.

* Check the quality and positioning of the wedges, which stiffen up the construction.
* Have a good look for what may be stored in this position. A stock of fuse wire left lying around may suggest electrical problems.

FIREPLACES

If there is an open fireplace, make sure that the flue is open by testing the draught. If the flue is open, remember when looking on the roof to check the chimney pots.

Where a gas fire has been installed consider the risk of the flue liner not terminating at the chimney pot.

See if the visible surface of the flue is sound. Weak and damaged linings increase the risk of fire. The chimney should have been regularly swept. Regular burning of wood may have deposited resin on the inside of the flue – this is flammable.

Replacement fireplaces, bigger and grander, require changes to the original chimney, which are often not undertaken in a sound manner. If the chimney breasts have been removed make sure that the stack has been supported. If the fireplaces have been removed, and the openings closed off, find out if the flue is open at the top.

FLOOR EXAMINATION CHECKLIST

PERFORMANCE

* Particleboard creaks – is it on insulation on beam and fill floor?
* Are the beams curved so that the insulation will not be able to sit flat?
* Edge support not provided in particleboard retrofit in old buildings.
* No insulation provided leaving hollow construction and noise penetration.
* Noise transfer where tiles used on concrete floor of kitchen or balcony.
* There is a risk of timber rot, partial collapse or beetle/termite infestation, where inadequate protection has been provided.
* For a hollow floor, is the level of ventilation adequate?

- How easy will it be to maintain built in pipes or other services.
- What service life do you expect based on the quality of installation of built in services?
- Electrics – can you amend the wiring without damage to the floor? E.g. if electric light switches were in the wrong place how problematic would a change be?
- Underfloor heating – cost or use of an electrical system, is it working, in need repair? Has the design allowed for the shrinkage of the floor construction or the ground beneath a heated floor?

DURABILITY

- Wear – what is the extent of risk resulting from the amount of wear that has taken place?
- Ability to clean – surface friction after cleaning, is there a risk? (e.g. tiles around a swimming pool or shower.)
- Dust – is it an indication of surface failure or screed failure?
- Tiles lifting – what method of adhesion, what levels of damp are there below the surface?

HEALTH

- Safety – slippery surfaces dry or wet, showers, rugs sliding on highly polished flooring.
- Steps – are there uniform risers and nosings?

MAINTENANCE

- Tiles – metric/imperial dimensions, ease of repair or replacement if damaged or sections removed for access to concealed services.
- T&G boarding or particleboard – what damage if they have to be lifted?

FINISHES

- Carpets – are they foam backed, and if so, have they powdered? (This increases the cost of their removal.)

* Are the carpets unsecured, how well is the stair carpet fixed, have the timber floor surfaces been damaged by heels, impact, furniture or marked by rubber soled shoes?
* How are carpets secured? Could they be removed, or are they a fixture? Do they form part of the sale? (They are usually a chattel.)

CLEANING

* Limitations – are they known?
* Polish – should certain types not be used?
* Water – what damage could be caused?
* Special cleaners – are they needed?
* Limitation of marble, unglazed tiles, etc. cleaning when stained.

SPECIAL NEEDS

* Wheel loading in warehousing.
* Hospitals – operating theatres hygiene.
* Skirting wall junctions where expansion provision needed, joint damp risk, services protection.
* Kitchens – public health requirement – cleaning washable/ movement of fittings.
* Computer access floors – wire and cables health and safety issues.
* Screed quality for support of flooring.
* Load storage, of what – paper and weight, eaves height.
* Children – any special need for impact absorption near equipment?
* Gym floor surfaces non-slip when moist.
* Wheelchairs – access for the disabled – shops/ public access buildings/ 2004.

Surveying Services

The cost of service installation in a new building may amount to 30% of the construction cost of a new office building, and a higher percentage of the operating cost for an office building. For a commercial building, those services may include the provision of a lift system, fan-coil air conditioning, fire alarm, security CCTV installation, telephone systems, IT installation, as well as the conventional electrical, gas, water and drainage systems, and lightning conductors.

An understanding of the service content of a building is important because a major part of the building, and a major cost in the management of the building, is involved with the services. The new changes in Part L of the Building Regulations seek to control carbon emissions created by building services. This will increase the cost of the external envelope and may reduce the overall cost of some of the services provision. The revised Part L will impose a range of requirements for the building services:

- Heating system efficiency is to be achieved by using more efficient boilers with enhanced controls.
- Hot water efficiency through discouraging local electrical hot water heaters.
- Improved insulation of building services.
- Improved lighting efficiency.
- Greater efficiency in mechanical ventilation and air-conditioning. The building fabric is not to add to the load.
- Energy metering.

The surveyor will find that there is a need for specialist assistance in the inspection of more complex installations, but should be familiar with the demands of each system, and be able to undertake a general level of appraisal in each case.

ELECTRICITY

Even though most informed observers recommend that most electrical installations be tested every five years, this rarely happens. Where, in the event of a failure in the electrical installation, there could be an even greater risk to life and limb, such as sports complexes, swimming pools, caravan sites, the storage of explosives materials, petrol stations or construction sites, there is a need for the testing to be carried out at much closer intervals, some on an annual basis. If such guidance were to be enforced it would mean that a purchaser should expect to see regular service inspection reports. Because so much of the wiring is concealed from view, no surveyor is going to be in a position to state that the installation is sound. In some cases the surveyor may see sufficient in the inspection to be able to report that rewiring will be required for the building.

IEE Regulations

The Institution of Electrical Engineers (IEE) set down the rules relating to the electrical installations within buildings in the UK. They cover requirements for the protection of householders or occupants of buildings against shock, and protection against the risk of fire being caused by overloads or faults in the wiring. Because these regulations require that the electrical installation be inspected and tested at least once every five years, it is reasonable to ask for the latest test certificate. If one is not to hand, or the last was carried out over five years prior to the inspection, an electrical test is required. At the moment there is no prescribed test method. Some insurance companies require regular testing as a condition for the continuation of their cover, and this is usually a specific requirement in old buildings.

The first electrical installation to a residential property took place in the 1920s following the creation of a reliable and cost-effective bulb in 1907. Up to the 1950s, most houses were wired in vulcanized rubber. The surveyor must be able to recognize acceptable wiring installations. The most common types of wire insulation are:

* Vulcanized rubber with cotton braid 1900–20. Now redundant.
* Lead sheathed cable 1910–39. Now likely to be faulty and will need replacing.
* Tough rubber insulated cable 1925–35. Round cable twin wires without earth.
* Vulcanized India rubber 1928–39. Often in conduit brittle without earth.
* Tough rubber sheathed (TRS) cables 1925–60. Vulnerable and redundant.
* PVC sheathed cables 1950 onwards. Older installations are redundant.

The introduction of square pin sockets coincided with the introduction of PVC installation cables. PVC is more durable, but will break down if hot cables are vibrated or the cable comes into contact with timber preservation chemicals or polystyrene.

Circuit breakers

In reporting upon an electrical installation, consider the fuse system or circuit breakers. The normal method of protecting against earth leakage is to connect the metal casing of a piece of electrical equipment directly to the earth. This enables faulty currents to flow to the earth, which will result in an overcurrent. Protection devices are installed within the installation to immediately terminate the supply. In older installations re-wirable fuses were used. This type of fuse is not acceptable. Fused systems should be upgraded because they react more slowly and are unable to detect current variations of less than one amp.

Miniature circuit breakers (MCBs)

MCBs can cut the supply if there is either an overload, short circuit or overheating in the system. The most sensitive type of circuit breaker is the residual current breaker (RCB). This is required by the IEE Regulations to protect anyone who may connect an electrical appliance for use outside a building. The cessation in current is achieved because the earth connector is given a very low resistance, and the current disconnects in a microsecond.

THE VISUAL INSPECTION OF AN ELECTRICAL INSTALLATION

Whilst it is recognized that the testing of an electrical installation is a specialist job, there are a number of defects in an installation that can be seen during a visual inspection.

+ Feel for warm socket outlets.
+ Feel for any warm connected plugs.
+ Look for burn or scorch marks around the socket face.
+ Look for cracked or damaged insulation to wires within the socket located in warmest part of building. Be careful in removing the socket cover.
+ Look in the roof space and examine the wiring:
  ◦ wires should not be placed beneath or within insulation;
  ◦ no wiring insulation should touch polystyrene granules;

◦ no exposed or inadequately protected wires should be visible outside junction boxes; and
◦ no cover of a junction box should be warm.

* Condemn old wiring serving as tails between meter and mains fuse.
* Condemn mains fuses of less than 1,000 amps.
* Condemn fabric covered wiring to any pendant drop.
* Listen for defective bearings in any fixed extract fan.
* Ensure that internal bathroom fans have 20 minute overrun after switch off.
* Note that night storage heaters are technically a fitting and could be removed by the vendor.
* Consider the adequacy of the number of sockets outlets.
* Check the position of sockets, no cable connetion to any fitting should need to exceed 2.5m (8ft).
* No power socket or light switch should be in a bathroom.
* No light or light bulb should be within reach of a tap.
* Shaver sockets should be equipped with an isolating transformer.
* Power sockets should not be within 3m of a shower located in a bedroom.
* Supplementary earth bonding must be connected to all metallic conductive parts in a bathroom, shower room, or a room containing a bath or shower.
* Supplementary earth bonding in a kitchen must link a metal sink with the water service pipes and any metal waste.
* A gas meter should have supplementary earth bonding within 600mm.
* Recessed downlighters must have fire protection if located within combustible flooring. Check the significance of any breach of the fire resistance of a floor separating different units of occupaton.
* Earthing must have an accessible but tamper proof terminal.

Test

Where a test is commissioned the examination should meet a specification that has been agreed before hand, that should include the following:

* Visual inspection.
* Confirmation that all equipment complies with the relevant British Standards.
* Continuity of ring final circuit conductors.
* Continuity of protective conductors and equipotential bonding.
* Earth electrode resistance.
* Insulation resistance.

- Polarity.
- Earth fault loop impedance – this should include an inspection of the location and ease of disconnection of the main bonding conductor. The test cannot be run without this disconnection.

POWER SUPPLY

For commercial office schemes, socket outlets are usually about one point per 5–10m^2 of floor space. For residential properties, the minimum provision will be about three double sockets for the reception room, dining room and the main bedroom, and two double outlets for each of the subordinate bedrooms. The kitchen should have switched spurs for each fridge, sink macerator, washing machine and dishwasher location, and three further double outlets.

LIGHTING

Within commercial premises the quality of the lighting must match the demand of the user. There are three categories of lighting fittings (luminaires) for use in areas where computer screens (Visual Display Terminals, or VDTs) are to be used.

Artificial lighting systems

Ceiling lighting is categorized by the visibility of glare during specific levels of computer operation. Reference within each category is made to the brightness of the luminaire being 200cd/m^2. This is dull. When looking at the light fittings at the varying angles of elevation to which reference is made, the surveyor should not be able to see the light source or lamp, only the diffuser.

Category 1 lighting should be selected where the dominant use of the space requires the intensive use of all of the computer screens all the working day. The light emitted, when measured at an angle of between 0–35° above the horizon, should not exceed 200cd/m^2. There will be no reflection observable when looking at a computer screen.

Category 2 lighting is required when a lot of terminals are in use, but they are not all in constant use. The brightness should not exceed 200cd/m^2 when measured at an angle of between 0–25° above the horizon.

Category 3 lighting is required if there is occasional use of a few terminals. The brightness is measured over the angle of 0–15° above the horizon. Some light reflection can be seen at the top of a screen.

LIGHTNING CONDUCTORS

The degree of protection required will be determined by the building's use and its construction. Maximum concern is given to schools, hospitals or places of public assembly, such as cinemas and theatres. In terms of construction, flammable external elements will expose the building to the greatest risk. Buildings with a metal face may induce a flash over, where the electrical charge runs across, for example, a roof, and discharges, not by the lightning conductor, but the remainder of the structure. Special steel or concrete framed structures offer a low risk.

An installation will start with a terminator at high level and a down conductor to carry the static charge to the ground. The earth strap must follow a simple route, and have a minimum number of bends. A single down service can serve a building of 100m^2. Where the down terminal runs across a metal roof, the roof should either be isolated, which is difficult, or be electrically bonded to the system. A testing joint should be provided to each down conductor.

ACCESS DUCTS AND FLOORS

For office accommodation, the relocation of staff and enclosures (churning) is a major cost. The surveyor should ensure that flexibility is provided in the design of the ducts, access ceiling voids and suspended floors. Ensure the requisite fire protection to ducts and access doors and that ventilation requirements have been provided.

Access floors

The depth below the access floor should not be less than 100mm at any point. The platform support must not damage the floor structure.

GAS INSTALLATION

There are several simple tests that can help a surveyor identify the extent of any visible defects. Testing will identify failures in concealed areas of an installation.

The following should be checked.

+ Gas pipes should be clearly marked.

- The pipes should only be coloured yellow.
- Mains connections should be either 20mm mild steel pipe or 22mm copper.
- Service pipe should not be routed through:
  ◦ unventilated voids; and
  ◦ under load-bearing footings or other foundations.
- Pipes under a footpath should be located to a depth of 600mm and not be closer than 0.96m from the kerb.
- Electrical earth bonding must only be fixed within 600mm of the supply-side of the gas meter.
- The gas meter should be located in a ventilated enclosure and be positioned not more than 2m from the entry point of the supply.

FLUES

- No opening or ventilation can be within 300mm of any burnt flue gases. Whilst this is the standard, flue gases do re-enter buildings with this minimal separation.
- Burnt flue gases will only be dissapated if the flue discharges into a well-ventilated area. Unburnt flue gases can suffocate the occupants of a building.

Open fire

- The flue for an open fire should be at least 200mm in diameter.
- A large open fire will require a chimney that is at least 15% of the area of the fire.
- Inadequate throating will increase the chance of smoke blowing back into the room.
- The height of the flue terminal will influence the effectiveness of the chimney.
- The best fireplaces will incorporate an air inlet from outside.
- The air supply into the room should be half the area of the flue at the throat entry to the chimney.

Gas fire flues

The terminal will depend upon the type of fire that has been installed. These are split into three categories:

- Decorative Gas Fires (DGF);
- Decorative Fuel Effect (DFE) fire; and
- Inset Live Fuel Effect fire (ILFE). This could have a small flue fitted which will be too small for use with a convential open fire.

Do not fit a gas terminal because this will reduce extraction of burnt gases. These gas fires are expensive to run and are inefficient. Some proprietary flue blocks have failed when they become hot. All flue pipes must be fully supported within a roof void and no angle change should be greater than 45°.

CENTRAL HEATING BOILERS

After March 2002 boilers must have a minimum SEDBUK rating of 85% for oil, 80% for LPG or 78% for mains gas.

* Check the adequacy of the ventilation provided. Boilers that are inadequately ventilated will endanger the life of the building's occupants and be expensive to operate.
* Poor flues will result in condensation forming on a mirror placed alongside the boiler. If this happens emergency work should be put in hand.

Condensing boilers

* The condensing boiler incorporates two heat exchangers. The first heat exchanger removes the heat from the hot gases in the same way as a conventional boiler. The gas then passes over a second heat exchanger that removes some of the latent heat.
* The flue gas is cooled so that water vapour condenses and is drained away.
* Large amounts of water vapour is discharged by way of the balanced flue.
* The boilers achieve over 80% efficiency.
* Problems with these boilers include:
  ◦ lack of reliability;
  ◦ poor hot water flow;
  ◦ risk of freezing of drain-off pipe;
  ◦ corrosive discharges; and
  ◦ the 'steam' cloud of flue discharge.

The surveyor's inspection should note:

* the type;
* age;
* capacity; and
* the service record of the boiler.

FLAME COLOUR

The colour of the gas flame is an indicator of the quality and efficiency of the boiler set-up, but to interpret the flame colour one must know which gas is being used – mains gas, or Liquid Propane Gas (LPG). LPG burns with an orange tip and is deep blue within the flame. Those colours in mains gas would suggest incomplete combustion.

Calorifier is a hot water storage tank and heat exchanger:

* Check the markings on the insulation to the wiring serving the immersion heater.
* Minimum insulation is 80mm thick pre-fitted shell.
* Multi-sectional jackets are not adequate.
* Ensure adequate capacity to allow for a reasonable speed of recovery.

To test that there are no problems at the time of the survey run your fingers carefully over the bottom edge of the cylinder to check for:

* failures that have been occurring due to excess water pressure;
* poor quality fittings; and
* corrosion.

SEALED CENTRAL HEATING SYSTEMS

Traditional central heating has an expansion pipe discharging over the header tank serving the system. In a sealed system special safety controls and an expansion tanks must be installed. The expansion tank allows the volume of water in the system to change as it is hot or cold. Air in the system is vented through automatic release valves. These must be kept unobstructed, unpainted and the system must be fully charged. The surveyor should check the condition of pipework in the usual way and make sure that the valves are maintained free and clear of paint.

MICROBORE SYSTEMS

In the 1970s a system was developed that used small bore flexible pipes linked to a distribution unit. The pipes are vulnerable to denting and being punctured. In hard water locations it is important that the water is treated to avoid furring. The boiler and pump must have been selected to work at the higher pressure required.

PUMPED WATER HEATING SYSTEM

This system relies upon a header tank servicing the water content of the installation. Two pipe circuits run around the building with radiators connected to both the flow and the return. Any loss of water is recharged from the header tank, and any boiler failure, overheated water or steam is vented back into the header tank. The header tank should be checked to ensure:

* there is water within the tank;
* the ball valve is opeartional; and
* the overflow is correctly connected and insulated.

WARM AIR CENTRAL HEATING

A central boiler serves warm air through a duct system. The warm air encourages draughts, is dusty and noisy to operate. The upper part of the building tends to be warmer than the lower. The air movement does reduce the risk of condensation, but unless there is a ventilated roof, the amount of moisture that condenses under the roof is much greater.

AIR-CONDITIONING

Air-conditioning is the control of the temperature, humidity and the quality of the air within a building, irrespective of the external weather. The term has come to include air cooling. Some installations referred to as air-conditioning only provide warmth or cooling. Concern about the level of fuel used in heating and cooling buildings has resulted in changes to Part L of the Building Regulations. New buildings will be expected to be designed to minimize the requirement for heating and cooling. The most common methods include:

- **Air handling systems** – require ductwork to allow air to be introduced and removed from each zone of the building. The trunking requires both horizontal and vertical space. This space demand will result in the building being 12% less efficient.
- **Direct Expansion (DX) Split Systems** – an outdoor unit (condenser) is linked to internal units. The indoor units can be above ceiling, below ceiling, cassette or wall mounted. Available in cooling or heating only, but provides a cost-effective solution, which is suitable for smaller installations.
- **Variable air volume (VAV)** – expensive to operate, maintain and install. It requires no water circulation. There are two types, standard and fan-assisted. In the standard system the total air volume is varied. In the fan-assisted form, the primary air volume is varied, with a constant air volume provided to the occupied zone. Adjustments in cooling are achieved by the increase in the supply quantity of air. Where a reduction in cooling is required, the supply of air is cut back. The minimum flow will be the fresh air needs of the occupants. This system's primary aim is cooling. To achieve heating a hot duct can be introduced into the air flow, whilst the cooling is operated at the lowest air supply level.
- **Variable refrigerant volume (VRV)** – involves an external condensing unit linked to indoor air-coil units. A condenser modulates the output to each condensing unit according to the demands of the zone

being served. It is the cheapest to install but the most expensive to run. The life expectancy of the main units is about 60% of that of other systems.

Fan coil systems

Usually located above ceilings but can be installed at low level around the perimeter of a building to provide heating and cooling. Four-pipe fan coils comprise filter, fan, hot water heating coil and chilled water-cooling coil. Two-pipe fan coil units provide cooling only and do not have a heating coil. Fan coil systems would normally be installed with a fresh air supply and extract system.

* **Four-pipe system** – the fan coil units have two coils, one for heating and one for cooling. Because the hot and cold water supplies are on different circuits the space occupier can determine independently the local space temperature. An auxiliary duct system is used to provide fresh air.
  ◇ The cold pipes must be insulated to a high standard.
  ◇ There is no control over the levels of humidity within each work space.
  ◇ The units are usually fitted to the perimeter to enable the inflow of air through the unit.
  ◇ The system requires a minimum duct provision and is suited to the refurbishment of properties where there is limited floor headroom.
* **Two-pipe system** – operates under the control of the building manager who decides whether coolant or heat should be provided. The unit has a single coil and filter.
* **Chilled beams or chilled ceiling panels** – an expensive system to install but the cheapest in the long-term. This passive method also reduces the disruption caused by fan driven air, reducing the need for cleaning or for filter replacement. Moulded panels with chilled water pipes attached are fixed or suspended from the ceiling. Ventilation, heating and air conditioning are dealt with separately. The movement of air over the panels will reduce the likelihood of condensation forming on the panels, and any decorations below being damaged. The system can cool where there is a small demand. The system takes up less space in the building and has the lowest operating and maintenance costs of all the cooling systems.
* **Under floor void** – the void beneath the access floor is used as trunking. The air is cooled by conditioned air modules (CAMs) that are connected directly into the floor void that is divided into supply and return plenums by aluminized fabric baffles. The inlet system

is individually controllable via a hinged panel in the supply grille or 'fantiles'.

* **Closed loop ground source heat pump system** – this is a method of acquiring heat for water or space heating. A small amount of heat is extracted by a heat pump from a non-toxic liquid circulated through pipework buried below ground. By compression in the heat pump, the temperature of the liquid from the pipe loop is raised and the heat is transferred, usually to the water in radiators or to heat domestic hot water. The pipe loops can be placed in wells, boreholes, lakes or below lawns.

DRAINAGE

Most drainage is concealed and even a visual inspection and testing cannot guarantee that no problems exist within the system. Some tests may damage the installation, and even where carried out their results have to be interpreted. Slight leakage is not unusual. The appearance of defects, particularly those shown on videos of the drain interior, do need to be put into perspective. The location and accessibility of the fault will affect the cost of repair. About 80% of the UK sewer network is rigid pipe work over 60 years old (20% over 100 years old), which is slowly breaking up. Data suggests that 40% of the existing network of rigid pipes exhibit cracking to some degree, with a possible 20% of new sewers in rigid materials exhibiting structural defects and cracks.

The purpose of the inspection of the drainage of a building is to record its present condition and location and to identify whether the system:

* has been correctly installed;
* is in working order and free from fractures;
* is free from blockages or any restriction;
* has a reasonable life expectancy; and
* is free from the need for major maintenance in the next five years.

Pipework

The type of pipe connected to the manhole should be checked if it is visible. Each pipe type will last for a different period of time.

* Cast iron is usually the best performer.
* Salt-glazed pipes can be durable, but is vulnerable to subsidence.
* Plastics and pitch fibre can have a short life.

Manhole

The examination of the manhole should reveal whether the system is running free. Any build up of solids within the chamber is probably the result of too slow a flow of water or a restriction in the drain. The quality of the sides of the manhole should be checked. Look for:

- Cracking in the render that may be caused by movement in the ground.
- The condition of the benching, fall not exceeding 1:12.
- The condition of any steps leading into the chamber.
- The quality of the drainage connection to the manhole.
- The pipe type of the drainage, and consider its life expectancy.
- The quality of the manhole cover.
- The security of the frame.
- The ventilation of the drain.
- Excess obnoxious smells escaping from the closed chamber.
- The possibility of the installation being a private sewer.
- The circumstances of the existing occupier and the possibility of blockage caused by: cooking with oil and fat; disposable nappies; or feminine hygiene products.
- The proximity of trees and the risk of roots affecting the drains.

No chamber

The absence of an inspection chamber must be drawn to the attention of the client. This is a serious limitation to any repair being effected quickly and economically if a blockage occurs. Where there is no manhole within the boundary of the property make sure that the deeds are checked to see if there is any shared rights and responsibilities of the parties to pay for repairs and maintenance to the drainage installation.

Interceptor manhole

This is the last manhole on the drain run before the connection to the street sewer.

Check for the following:

- The presence of the interceptor stopper above the water trap.
- The air vent serving this manhole.
- The presence of any blockage to the trap.
- The quality of manhole cover.
- The security of the frame.

Preferred method of drainage in new work

Order of priority for rainwater disposal (Part H Building Regulations)

(1) A soakaway
(2) A water course
(3) A sewer

Order of priority for foul water disposal

(1) A public sewer
(2) A private sewer
(3) A septic tank or treatment system
(4) A cesspool.

Ventilation

The condition of the mica flag in the fresh air vent should be inspected. This allows air into the drainage system so that it can escape through the vent pipe.

Rats

The problem of rats entering the building's drainage via broken or defective rigid pipework is significant. The failure of flexible pipe joints and pitch fibre pipes have all contributed to their escape. *Rattus Norvegicus*, the brown rat, is currently the dominant common rat species in the UK. It is also called the black, house, sewer or wharf rat.

NON-MAINS DRAINS SYSTEMS

*   **Cesspool** – a tank into which all effluent and grey water flows. Emptying is undertaken for a charge. Tank capacity should be at least $18m^3$, which will need emptying every 45 days. Tanks should not be located more than 15m from a dwelling or 30m from the access point for emptying.
*   **Septic tank** – a system for the treatment of effluent and grey water and the discharge of liquid into the surrounding ground. The treatment involves the passage of the waste products into a chamber where the solid content settles to the bottom and reacts with anaerobic bacteria that exist where there is little or no oxygen. The run off then flows through a chamber where the liquid is exposed to aerobic bacteria. The system requires sufficient ground to be able to soak up the discharge. Clay ground is not suitable. The system requires 'desludging' every few years depending upon its usage.

TESTING DRAINS

The drainage system can be tested in the following ways.

* **Water pressure test** – (hydraulic) the system is blocked at the lower end and filled with water. The water level within the pipes should remain reasonably constant over a test period of approximately half an hour. Where a failure occurs, there will be a need to test branches individually to find the faulty section. There may be some absorbtion into the brickwork of the manhole. The test places greater pressure on the lower part of the system.
* **Air pressure test** – (pneumatic) the drain is sealed at both ends and air is pumped into the pipes, producing an even pressure on all the pipework. The pressure is checked to ensure that there is no seepage, or leak, from the system. Where testing between a sealed point and a sanitary fitting, the water trap will fail at very low pressure.
* **Smoke test** – used to trace pipe runs and connections, but does not test for faults in a drainage system, unless linked with an air pressure test. Mainly used to test pipes above ground where smoke escape from joints would be seen.
* **Video recording** – cameras are run along pipes recording the appearance of the pipe interior. Work should be supervised, to ensure quality of information. Restrictions prevent ongoing examination. Cracks may not be defects. This gives an accurate location of those defects found, its value is often oversated.

ABOVE GROUND DRAINAGE

* **Rodding access** – all pipes must be capable of being cleared if any blockage does occur. Rodding eyes must be accessible.
* **Gulleys** – rainwater pipes should discharge either into a back inlet gulley, or discharge over a grid that connects to a trapped connection to the drain. The trap will collect any grit washed from the roof and should be cleaned out at regular intervals. Open up the trap and test it by filling it with water. It should hold the water above the drain connection.
* **Pipe supports** – should be provided at regular intervals; for horizontal pipes of 50mm diameter every 0.6m, 100mm diameter every 0.9m, and vertically at 1.2m and 1.8m.
* **Waste pipes** – should be ventilated at the head of the system above the highest window.
* **Pipe movement** – plastic pipes can move 3mm/m. Ensure adequate provision has been made in jointing.

* **Fire protection between floors** – check passage of plastic wastes through fire barriers.

SANITARY FITTINGS

Water leaks from any sanitary fitting or appliance may have serious consequences for the remainder of the building. During the inspection check, by rubbing joints with the hand:

* for leaks in hot and cold connections to all fittings when accessible;
* for leaks in water connections to washing machines and dish washers;
* for leaks from all waste connections to all appliances; and
* for leaks to all piped connections to heating appliances.

Test the controls and wastes as follows:

* Find the main stopcock, ensure that it is capable of use and readily accessible.
* Test each tap to ensure it is connected, for water flow and mains pressure.
* Test, where operating, that there is adequate hot water of an appropriate temperature.
* Check all mixed material instalations. There can be an electrolytic reaction if galvanized steel and copper pipes are in contact.
* In a hard water areas, there is a risk of lime deposits restricting the pipe bore.
* Check all pipe insulation for the risk that it may contain asbestos.
* Check all pipe insulation for adequacy.
* Test the performance of the waste to fittings by filling the bath or basin, letting out the water, and listening to the sound of the last part of the discharge. Pulled traps create reactions in connected fittings. Check the depth of the remaining water in the trap.
* Lift the cover to the Waste Water Preventer (WWP), which must be accessible. The ball valve should respond to a finger touch, and turn off when the finger releases. Check the connections to the overflow pipe.
* Examine the floor around the shower tray and test the walls to each side for dampness.
* Look at the shower tray and see whether it is made from ceramic (less usual in modern work), fibreglass or epoxy resin. The resin trays should be bedded in cement on a firm base, and will sound solid when tapped.
* The bath holds a lot of water, and its weight with body and water may have damaged a weak floor. Check for signs of movement.

SURROUNDINGS

- **Boundary fences** – there is no rule that says one must fence one's land, unless there is a covenant in the lease or an agreement that could be enforced. Where fences do exist, their condition will have a cost implication.
- **Retaining walls** – where there are varying ground levels the state of the boundary walls must be verified. Some household insurance policies do not include the retaining walls within the household cover, and the cost of their maintenance and repair may be substantial. A careful check as to the condition and a warning as to the implication of the repair cost is an important part of the survey report.
- **Paths** – note their condition, consider the implications of a failure. If the ground has been eroded by the failure of pipes or drains, some irregularity in a lawn or path may suggest the point of failure.
- **Paths safety issues** – are they slippery, uneven or have steps that could cause an accident? Are there adequate handrails where the path may be uneven or run close to a drop?
- **Garages and sheds** – all sheds and garages should be looked at, even if they are not to be part of the report and may be excluded from your liability. Check whether they are a chattel, being capable of removal or are a permanent fixture. Their condition may suggest a risk of early failure in the main building if built with the same materials. Movement in the ground may have affected the garage first and be going on to cause problems for the main house in later years. Look at the contents to see if they suggest what recent maintenance has taken place.
- **Flooding** – the risk of flooding will have been checked before the inspection, but the presence of sandbags tucked away behind a shed or garage may underline the risk present. Where would the excess water of heavy rainfall run to in a thunderstorm?
- **The drive** – there should be a proper pavement cross-over that provides adequate sight lines.
- **Security** – are there security implications for the exterior of the building? The climbability of fences or barriers may warrant consideration. Identify the absence of locking bolts on windows or a decent lock on external doors and look for signs of broken car windows in street parked vehicles.
- **Service access** – if the property is served by overhead wires, be they power or telephone, see if they could be damaged by gale-tossed trees.
- **Fuel tanks** – if there are fuel tanks within the grounds, be they oil or gas, how easy is the access for their replenishment?

* **The street** – has it been adopted? If it hasn't, there may be a future and substantial responsibility for the building owner to pay towards the cost of maintenace or the upgrade of the road for adoption.
* **Other properties in the road** – carefully look at the other houses in the street, to see if they suggest weaknesses in that building type. They may confirm the life expectancy of the roofing, the durability of window frames, the status of a neighbourhood, the risk of vandalism and theft within the area.
* **The lessons of the street** – does the condition of other properties in the street suggest a weakness in the building being inspected? Are other houses in the street being re-roofed, or having underpinning work carried out? Have they all had replacement windows and gutters? A careful look at the lessons that can be learnt from other dwellings may assist in the final checks that you make to ensure you have missed nothing in your inspection.
* **The door knob test** – before you leave, are you satisfied that you understand the condition of the building. Are you satisfied with the work that you have done? Is there part of the building that is nagging you, where you are unsure about the level of risk? If so, go back.

Dampness

Dampness is inextricably linked to most building deterioration. Water contributes to the oxidization of metal leading to the corrosion of steel reinforcement in concrete and the propagation of fungal decay or beetle infestation in timber. Water also facilitates chemical changes in the components of a building as well as being a source of damage where it penetrates into a building. For these reasons the inspection of a building must take into consideration how dampness may enter the building as well as identify the damage that it may have caused.

Although the back of your hand may help you decide if excess moisture is present, there is no way that the precise percentage of moisture can be determined without an accurate moisture meter, and then most can only achieve a reliable reading in timber. Readings in plaster, concrete and brickwork are a guide and not readings of the percentage moisture content. The smell of any decomposition caused by the dampness, or the sight of damage caused by dampness, will assist in identifying that dampness is present.

Knowledge of the exact moisture content will be required for one to advise whether an increase or decrease in moisture has taken place or whether timbers may be at risk of deterioration. The moisture meter will assist in identifying the pattern of moisture within a building. Even if it may not be accurate, it will show if one part is damper than another. With a resistance type meter, the observation, test and reading are quick.

The moisture content is the amount of water in a material divided by the weight of the material. This means that a heavy material would have a lower percentage moisture content than a light material for the same volume of water. For example, 5% moisture in mortar is dry, in brick it would be damp and in plaster it would be saturated. Plaster would be regarded as being wet if it had a moisture content of over 1%. For a brick, dampness would be identified if the reading were above 3% and for timber over 20%.

Dampness in conventional walling materials has been given a watershed figure. If any component of a wall, be it the mortar, brick or plaster, is in excess of a 5% moisture content, it is to be assumed that the wall is damp. There is an ambiguity around that 5% because wall or floor components with a moisture content in excess of 5% may require no action as the moisture content is within a tolerance limit, whereas defects recording a moisture content below 5% may require urgent attention.

It will be reasonable to say that dampness occurs where a material is wetter than air dry. 'Air dry' means in equilibrium with a typical atmosphere with a relative humidity of between 40–65%. The agents of decay in building materials – moulds, decay fungi, mites – have a biological origin. They are able to develop where the relative humidity is between 75–85%. (Timber in such air conditions would have a moisture content of between 17–20%.)

The relative readings on a conductance type electrical moisture meter indicate the relative dampness of different materials because they measure the free water in a material. These measurements are a fairly close representation of the relative humidity. A high reading on such a meter would indicate a damp condition of approximately equal significance in wood, brick, plaster or wall board regardless of their very different moisture contents. It is for this reason that these meters are usually marked in zones indicating safe, intermediate risk, and danger of damage or decay occurring.

REPORTING DAMPNESS

Dampness is an elemental part of most survey reports. It is a targeted part of the HOMEBUYER Survey and Valuation Report format, which refers to the identification and reporting of dampness as being one of the four major defects that are to be emphasized within the report (the others are movement, timber defects, and insulation).

Knowledge of dampness

A surveyor, undertaking any form of building inspections, should be aware of the published information on dampness. For example, BRE Digest No 297 (published May 1985) states that: 'The selection of an effective remedy for any dampness problem must start with a correct diagnosis of the cause'. BRE Information Paper IP19/88, 'House Inspections for

Dampness', gives guidance as to the areas of a building where there is a risk of water penetration.

The surveyor should also be aware of the results of research into the extent of rising damp and the ease of misinterpreting signs of dampness to low level within buildings. In April 1997, the BRE issued 'Diagnosing the Causes of Dampness', a repair guide for building owners and managers that looked at the problem of rising damp. In that document they advised that:

> surveyors are often too quick to assume that damp problems are caused by rising damp. As true rising damp is, in fact, not very common and rising damp remedies can be very expensive, it is important to correctly diagnose the cause of damp problems.

The BRE have advised that rising damp was found in only 10% of the buildings claimed to have rising damp. Water is reluctant to rise within brickwork in an independent wall, but water within the walls of a building will react to the dynamics of the construction.

- Walls that are encased, e.g. in damp retardant plaster, in linings or claddings, can react to pressure within the wall and create water movement above 1m.
- The installation of a solid concrete ground floor with a damp membrane linked to wall treatment encourages dampness to rise in enclosing walls.
- The movement of water, and the pressures that are associated with different wall compositions, may encourage dynamic forces to move water within building construction.
- Salts deposited by the evaporation of moisture within a wall can encourage the wall surface to attract water from the humidity in the air. A modern damp meter will read dampness where these salts have been deposited, and that has confused some diagnostic work.

A surveyor, undertaking this type of survey work, must be familiar with the limitations of a moisture meter. It can be used to check for dampness within a building. The surveyor is expected to be aware of both the limitation of the meter's readings and the need to interpret the information that such a meter will provide.

Rising damp exists

Rising dampness is 'an increase in water content in a building component caused, in whole or in part, by the redistribution of moisture from a lower level'.

Rising dampness may not have risen through the wall from ground level as is usually supposed. It could have entered the wall above ground level, by plumbing failure, by damp courses being bridged, by transverse penetration from adjoining construction, or by water seepage due to leaks from roofs, gutters and down pipes.

The failure of a damp course is only one of many possible causes. Dampness within a wall is often the result of more complex causes than the movement of water through small gaps within the lower part of the wall. Some part of the walling will be porous if rising dampness has taken place.

Dynamic movement of water can result from:

* Different environments existing on each side of it, e.g. different temperatures.
* Salt from ground water. This can attract free water from the air. Warm moving air will dry the moisture from one side of the wall, leaving the salts on the surface.
* Water can be extracted from the air by condensation or salts, enter the wall and move up to evaporate above.
* Salt deposited on or close to the surface could seal the wall surface. This can encourage water sourced from the ground to move even higher up the wall.
* Varied water pressure, for example, by variations in the water table.

Rising dampness can be caused by a mixture of events and circumstances.

* The variation of any of the static influences or the dynamic variations in circumstances will increase or decrease the height to which the water will climb, within limits.
* It will also influence the volume of water that may move through the construction of the building.
* The level of the water table.
* The resistance of the wall surface to evaporation.
* This damp barrier or damp course has the aim of preventing water moving upwards through the wall.
* Deterioration of the damp membrane or damp course.
* Penetration of dampness from the lower part of the wall.

Damp courses

Materials that have been used vary from:

- Engineering bricks.
- Slate.
- Lead.
- Polythene.
- Bitumen-impregnated felts were introduced in the 1950s onward.

Common causes of damp course failure:

- Structural movement fracturing brittle damp course.
- Light degradation of polythene membranes during construction.
- Design error in cavity trays or stepped damp courses.
- Incorrect ground levels too close to damp course.
- Bricks laid frog up and no mortar bed below the damp course.

Repair options

Remedial work must correct the original cause, and not be selected just because dampness exists within a wall.

- **Chemical injected damp courses** – rarely effective in old buildings. Essential that internal plaster replaced to a height of 1.2m. There is doubt whether the injection is able to replace the moisture that may already be within the wall if the solvents do not enter those fissures that are retaining water. That may result in the fluid advancing further as fingers than as a united front.
- **Physical barrier** – introduced into walls by horizontal saw-cut. Access must be available to both sides of the wall. This membrane needed to be laid between mortar beds, so that it was protected. Disruptive but effective repair.
- **The Massari method** – developed in the 1980s. The system involves the introduction of ceramic angled tubes into the outside walls of the building, which are supposed to enable any moisture within the wall to evaporate. Not universally accepted as being successful.
- **Electro-osmotic damp course** – used from the 1970s, but have fallen out of favour. The existence of an electrical current can encourage water to rise within the wall. The systems were either active or passive. The active system works with the placing of a direct current through the wall with the intention that the water particles are driven from the anode to the cathode. The passive system relies upon the connection of the electrode to the earth, there being sufficient water in the wall to complete the circuit.

- **Internal replastering** – most installers will emphasize that a replacement damp course is part of a two-part process, the second part being the replacement of the internal wall plaster. If the plaster is replaced at the same time as the damp course is replaced, there is the chance of salts being drawn to the plaster face in the drying out process.

Condensation

With very rare exceptions air always carries moisture. Warm air is capable of carrying a larger amount of moisture than cool air. The temperature at which saturation takes place is referred to as being the 'dew point'. If this air meets a surface that is at a lower temperature than the air temperature, the moisture in the air will condense into water upon the cooler surface, thus forming condensation.

The standard of insulation within buildings has increased regularly since the mid-1980s. The changes to Part L of the Building Regulations in 2001, further raised this standard to create efficient energy saving walls, floors and roofing to new buildings. These changes have reduced the amount of natural ventilation that is both provided and which may exist within a building. New homes may achieve little more than three changes per hour, when buildings only a few years old may have achieved 14 air changes an hour.

The creation of walls that no longer have cold areas will result in more condensation forming where air movement is minimal. There has been no medical study that states the minimum number of air changes that are required for healthy living. The health risk will vary dependent upon the age and health of the occupants.

In poorly heated houses with little ventilation, the humidity of the air will tend to be quite high because the moisture created by human activity is not dispersed through the intake of less humid air. Condensation will occur on windows, and on wall surfaces, and mould growth will be seen on carpets, furniture and on stored clothes.

Contributors to condensation:

- Human beings.
- The use of gas and paraffin heaters.
- Furniture placed against outside walls.
- The absence of extract fans in a kitchen.
- Bathing or showering with an open door and no ventilation to remove the moist air.

- Boiling food.
- Heating systems that raise and lower internal temperatures.
- Variable wall temperatures or cold bridging.

Mould growth is common in buildings that have condensation problems. In order to avoid mould the relative humidity should be kept below 70%. Reducing the moisture content of the air can reduce the risk of mould growth, as can increasing the level of heating, the amount of ventilation and the insulation to the building fabric. The mould starts as individual spots and develops into larger patches. It occurs in varying colours, usually green, black or brown.

Roof void

In shallow pitched roofs, with a roof slope of less than 20°, there is a possibility that moisture may build up against the timber components of the roof. The lower the slope, the greater the obstruction to through ventilation, and the greater the risk of inadequate discharge of the moisture from within the construction.

Most flat roofs do not have adequate through ventilation. Any moist air that gets into this cold roof will condense below the roof finish. Where the flat roof is of concrete construction the colder surface of the concrete usually causes condensation to occur on the ceiling.

THE SURVEYOR'S ADVICE ABOUT DAMPNESS

Dampness at ground level must be interpreted. To make a diagnosis the surveyor must:

- examine the construction;
- identify the risk of failure;
- identify the extent and location of the dampness;
- consider the chance of a damp proof courses being required or damaged;
- look for alternative causes; and
- recognize there may be more than a single cause.

Where to look:

- The ground floor junction with the walls throughout the building.
- Around the windows in the building.
- The internal surfaces behind any gutter or downpipe.
- The junction of any roof to a wall.
- The base of parapet walls.

Table: Diagnosis

Circumstances	Rising damp	Condensation	Penetration	Pipe leaks
Water on wall face when wiped with one's hand	Unlikely – almost never happens	Possible – the surface when wiped often feels greasy, but this will depend upon the extent of condensation and the surface porosity. Where there is an impervious surface, wall tiles, or a vinyl paper, surface water will occur	Possible – depends upon the extent of penetration	Probable – depends upon the extent of water escape
Crunchy salts on the wall surface, often behind the wall covering	Probable – the salts are deposited as the rising dampness evaporates, leaving the salts in the surface of the wall or the plaster	Very unlikely	Very unlikely	Very unlikely
Mould growth	Unlikely	Very likely – patches of mould in poorly ventilated locations	Possible – it depends upon the extent of water penetration	Possible

Moisture above 1m	Possible – the usual pattern for moisture readings will show a decline from around 300mm and an absolute line where no reading is obtained one side and a measurable level on the other side	Possible – does depend upon the conditions of the room and the source of water supply plus the temperatures of the wall	Possible – depends upon the location of the point of penetration	Possible – most pipes are located at floor level, so that seepage is usually at low level. The flow down into rooms below can cause high-level dampness
Moisture readings in timber	High readings usually obtained	Low readings are usual	High readings if water penetration is at low level	Water pipes being at low level, it is usual for the moisture readings at low level to be very high
Moisture within plaster	Damp readings should be present through thickness of the wall	Damp readings more common at the face and declines as one progresses into the wall. The walling should not show moisture content at its core	Usual to find isolated instances of water entry, with the wall damp in a profile that can identify the epicentre of the water entry	Usual for the wall dampness to be on the surface, unless the water escape has been held on top of a damp course so that the wall is able to soak up the dampness in the same way as rising dampness

Timber Decay Caused by Insect Attack

IDENTIFYING THE INSECT

The surveyor needs to be able to identify which beetle has caused the timber damage, because this may help identify the extent of damage that may exist. Most timber damage lies inside the wood, concealed from a surface inspection. The visible holes are the exit holes usually of the male insect; the female often remains within the timber, producing grubs to increase the amount of damage. The deathwatch beetle causes extensive damage in hard wood where it is bedded into the wall, possibly because the wood is a little moist at that point. Even stabbing the wood may not discover the extent of timber damage, or the prospect of dry rot, which is frequently associated with an infestation.

Risks

* Treatments undertaken in the 1970s are becoming less resistant to insect attack. Strains of insects have developed with a resistance to the usual control chemicals.
* We have imported insects in timber.
* Central heating does not eliminate the risk of insect attack. Most conventionally constructed buildings hold above 12% moisture in the sub-floor, roof voids, under-stairs, and wall mounted cupboards for beetles to enjoy their wood with a little gravy.

Means of recognition

* To enable an accurate identification, one needs a magnifying glass. My favourite magnifier is a 'Loupe',

a fixed focus lens, which enables examination of any beetle retained within the Perspex cup beneath.

* The destruction of wood is carried out sometimes by the grub, and sometimes by the adult form.
* The manner of the spread of the attack varies with different insects. Some will attack only the living tree; some will only eat decayed wood, or timbers close to the bark.
* Some insects require warmer climates.
* In a temperate climate, some insects can multiply rapidly.

TREATMENT

The treatment of insect attack may be carried out in three ways.

* The use of high temperatures.
* Suffocation – by blocking the pores in the insect's outer covering.
* Poison – timber is impregnated with tasteless preservatives containing arsenic or lead salts which enter the stomach of the insect when the wood is eaten.

Concerns over timber treatment

* The toxic effect on human beings.
* The effective dose which will destroy the infestation.
* The risk to the building if the treatment is not successful.

Continued overleaf

Table: Timber decay caused by insect attack

Insect/beetle	Length (mm)	Flight hole (mm)	Shape	Treatment	Wood damage
Bark borer	4–6	1–2	Round	None – remove bark	Bark
Biscuit beetle	3–5	1–2	Round	Dispose infected material	Nil
Bostrychid Powderpost	4–11	3–6	Round	None – dies out in 12 months	Irregular tunnels
Carpet beetle	(10mm)	3–6	Round	Dispose infected material	Nil
Common furniture	3–5	1–2	Round	Organic solvent, emulsion or paste	Sapwood, old furniture, loft timbers, under stairs
Death watch	6–9	2–3	Round	Organic solvent or paste	Hardwoods, oak and elm where slight dampness
House longhorn	10–20	6–10	Oval	Organic solvent or paste	Sapwood of softwoods, particularly roof timbers
Marine borer	>300mm	9	Round	Dry timber	Immersed wood

Pinhole borer	3 (up to 6)	1–2	Round	Dry timber	Standing only
Silver fish	12			Remove damp	Nil
Weevil	3–5	1	Round	Dry timber	Damp wood only
Wood lice	15			Remove dampness	Nil
Wood wasp, masonry bee	18–35	6–8	Round	None – exits within 12 months	Newly felled or diseased trees

Rot and Fungus Damage to Timber

DATING AN OUTBREAK OF DRY ROT

* It is not possible to accurately date an outbreak from a visual inspection of the timber.
* The date of the cause of the dampness is a better guide.
* An examination of the fungus may assist in giving some general date of the start of an attack.
* The speed of spread may be of assistance, but this will vary depending upon the moisture present, the type of wood and the temperature and air movement.
* The prediction of the spread of dry rot is not exact. The speed of spread will vary. In optimum laboratory conditions on plate cultures it can be between 4mm and 2.25mm a day.
* The average growth rate is thought to be about 1m per year.
* In ideal circumstances the spread could be faster.
* The depth of the cracking in the timber may help in estimating the age of an attack. In new, open-textured timbers, dry rot can penetrate to a depth of up to 18mm in a year, whilst in older, denser wood the penetration is only a quarter of that amount.
* It is common to find outbreaks that have died or become dormant. One may assume that the source of water dried up. Some believe that such an outbreak is merely dormant, awaiting a return of water for the infestation to resurrect itself; whilst others believe that the fungus is dead.

Future dry rot treatment?

Experiments are being carried out where the amount of ventilation is increased to see if this can control and eliminate existing attacks in old buildings. If this does work, it will allow more original features in infected buildings to be maintained than has been the case before now.

Risk of an outbreak

* There is a high risk for buildings in an exposed location, higher still if walls constructed using non-absorbent materials, such as granite.

- Anywhere that has a concentration of water flow soaking into the mortar bed between brick or stone.
- Stone buildings where render has been removed from external wall faces.
- Where there is a high risk of water escaping within the building.
- Do-it-yourself plumber.
- Heating systems where the pipes are run in the floor screed.
- Buildings that have had past infestations.
- Buildings that have suffered from settlement or subsidence where drains may have been damaged, damp courses punctured or broken, and water services distorted or failed.
- Inadequately ventilated flat roofs.

Freshly cut timber has a moisture content of between 40–90% of its dry weight. Much of this moisture is lost soon after felling, because it is mainly the sap water held in the timber. The remaining moisture is held within the fibres of the timber and this is somewhere between 25–30% of the dry weight of the timber. Further reductions are only obtained by seasoning the wood. Kiln drying means that most timbers will have a moisture content in the region of 7–10%.

IDENTIFICATION OF THE TYPE OF ROT

Identification of the type of fungus cannot be achieved by inspection of the hyphae alone, a thread of narrow tubes or filaments which grow by extending their tips and branching frequently. Sectional analysis of the hyphae of different fungi does not reveal sufficient variation to accurately identify each species. The identification of the type of disease has to be achieved by an examination of all the characteristics of each fungus.

As the hyphae grow they form a mat of material known as the mycelium. Most of the various forms of fungus have individual characteristics in the mycelium. The fungi reproduce by spores which are distributed from the reproductive structures known as fruiting bodies. These spores are readily dispersed by the air. It is believed that the spores carry a small food reserve, so when the spore settles it must find food quickly or perish.

SERPULA LACRYMANS

DRY ROT

Decayed wood tends to develop a cuboidal cracking formation to the surface of the timber. The moisture

Table: Resistance of home-grown hardwoods to decay

Timber	Common Name	Average loss in dry weight (as percentage of original dry weight) caused by test fungi after four months at 22°C				Class of decay Resistance
		Serpula lacrymans	Coniophora puteana	Fibroporia vaillantii		
Acer pseudoplantus L.	Acer sycamore	26.6	20.8	–		Perishable
Aesculus hippocastanum L.	Horse chestnut	33.6	29.9	27.3		Perishable
Carpinus betulus L.	Hornbeam	21.4	11.8	9.4		Perishable
Castanea sativa Mill	Sweet chestnut	Negligible	Nil	Negligible		Resistant
Fagus sylvatica L. (two tests)	Common beech	26.2	30.9	13.2		Not resistant to perishable
		39.5	36.5			
Ilex aquifolium L.	Holly	3.1	3.0	2.9		Not resistant
Jaglans regia L.	Walnut	3.4	5.1	9.3		Moderately resistant
Pyrus torminalis Ehr	Pear	11.9	24.1	–		Perishable
Quercus robur L.	Oak	Negligible	4.0	–		Resistant
Ulmus hollandica Miller	Dutch elm	16.0	–	17.2		Not resistant
Ulmus procera Salisb	English elm	15.5	21.6	1.9		Not resistant

content required for propagation is between 20–35%. Rapid growth can take place with temperatures as low as 5°C. Growth stops at 0°C and above 26°C. It is remarkably sensitive to heat and will die when exposed for a very short period to temperatures of 40°C.

Its rate of growth, given reasonable conditions, is about 1m per year. Most active growth occurs in conditions of bad ventilation and high humidity. The fungus mycelium develops on the surface of infested timber and may take the form of a soft white cushion with a cotton wool texture. Under dryer conditions, a skin of silver grey silky mycelium, sometimes with yellow patches and tinges of lilac, may be formed on the wood surface. Droplets of water are produced at the tips of the hyphae and lie on the surface of the mycelium. Feeder strands develop within the mycelium and supply water and nutrients to the growing area. These strands may extend for several feet over brick or steel. Eradication requires the killing of the feeder strands.

The fruit bodies (or sporophores) are fleshy growths resembling a flat plate or bracket. They are initially pale grey, tinged with yellow, but as the spores develop the spore bearing surface becomes rusty red in colour. This surface may be corrugated into irregular folds (pores) and has a characteristic mushroom-like smell. The fruiting body discharges its spores up to 800 million a day for up to two weeks. The fruiting body then shrivels to a dark brown colour and dies.

Surface damage	Cuboidal cracking and darkening of the wood. Note deep cross-cracking
Identification	Strands grey, become brittle when dried. Silver grey growth similar to cotton wool, slight yellow tinges to edges. Fruiting body, soft fleshy labyrinthine, white edged, rusty middle
Rate of growth	1m a year for the strands, although flash growth of the mycelium can be greater
Light	Reactive to light. Required for growth of fruiting body
Dampness	20–35% (i.e. damp, not wet)
Temperature	0–26°C. Optimum growth at 22°C. Will die if exposed to temperature of 40°C

CONIOPHORA PUTEANA

WET ROT

Thrives in wetter conditions than those required by dry rot. It can also be found in worked timbers and logs in the open air. *Coniophora puteana* is highly vulnerable to fluctuations in moisture content and achieves optimum growth in timbers with a moisture content of between 50–60%. It is unable to survive if the timber's moisture drops below 43%.

Surface damage	Cracks follow the line of the grain. Minor cracks only across the grain
Identification	Thread-like strands, yellowish, becoming darker brown with age
Fruiting body	Flat sheet-like with olive brown spore knobbly surface
Rate of growth	Up to 40% weight loss in four months
Light	No recorded reaction, usually found out of light source
Dampness	Over 35% but limited life expectancy above 49% (very wet)
Temperature	Optimum 23°C

FIBRIOPORIA VAILLANTII

MINE FUNGUS

Although this is a common fungus in mines, it sometimes does occur in very damp timbers in buildings. It can be confused with dry rot. It has a white mycelium but does not show the traditional lilac or yellowish edges which are common with *Serpula lacrymans*. When freshly growing, the surface of the mycelium is fern-like. The strands are smaller than those of dry rot but remain flexible when dry. It is less likely to have penetrated loose brickwork or lime mortar than dry rot.

Surface damage	Cuboid cracking in the timber when dry. Darkens the wood
Identification	Strands whitish and seldom thicker than strong twine, remaining flexible when dried
Fruiting body	Not usually in buildings, plate shaped with honeycomb surface, white in colour. Associated with water leaks. Will not spread to drier parts of building
Rate of growth	1.4m a year
Light	Not usually found in light
Dampness	45–60%.
Temperature	Up to 36°C. Optimum 27°C

PAXILLUS PANUOIDES

This occurs in softwood timbers which have been allowed to become wet and is also common in damp underground conditions such as mines. It is able to propagate outdoors in softwood tree stumps or sawdust.

Surface damage	Wood becomes soft and cheesy and on drying out has deep cross cracks. Surface damage occurs in softwood
Identification	Brownish or dingy yellow shell or bell-shaped fruiting body. Mycelium yellow with occasional violet patches. It causes a yellow discolouration of wood
Rate of growth	Up to 0.5m per year.
Light	Will propagate in daylight
Dampness	50–70%.
Temperature	5–29°C

DONKIOPORIA EXPANSA

In the UK it is largely restricted to attacks on oak and chestnut inside buildings and tends only to propagate in warm areas where there is an absence of light.

Surface damage	Reduces wood to fibrous strands
Identification	Fairly common on damp large oak timbers. Large plate-shaped fruiting body up to 200mm long and 25mm thick, often in several layers. The fruiting body is hard and woody, warm buff to brown in colour. Causes an active white stumpy rot. Hyphae are yellow and are plentiful in wood
Can cause	35% weight loss in four months. Linked with death-watch beetle
Rate of growth	Rapid, but will not extend beyond timber
Light	Absence of light required
Dampness	20–36%
Temperature	Optimum 27°C. Maximum 35°C

LENTINUS LEPIDEUS

STAG'S HORN FUNGUS

Not a common fungus within the UK – it is found occasionally in the stumps of conifer trees.

Surface damage	Leaves wood in brown cubical shapes
Identification	Occurs in telegraph poles, railway sleepers, paving blocks, and occasionally in flat roofs. Distinctive fruiting body of either antler-like development of thin mushroom with extended stalk. Colour varies from pale brown to a purplish brown. Tolerant to creosote except in very high quantities. Strong smell like balsam
Rate of growth	Up to 1.5m per year
Light	Not affected by the presence of light
Dampness	26–44%
Temperature	Optimum 27°C. Maximum 37°C

DAEDALIA QUERCINA

Found in oak in UK, but also attacks chestnut, beech and some other species. In buildings it is most likely to attack timber windowsills or doorsills where rainwater has not drained away. It tends to propagate more easily in the warmer south-west of the UK.

Surface damage	Reddish-brown cuboidal rot. The individual cubes are often quite large.
Identification	Thin sheets of mycelium in spores in wood. The decayed wood is friable. It attacks hardwoods, usually oak, and it smells of apples
Rate of growth	Up to 0.8m per year
Light	Cultures tend to be more silky in the dark
Dampness	Optimum 40%
Temperature	Optimum 23°C. Maximum 30°C

PHELLINUS CONTIGUUS

This fungus occurs on wood which has been worked. It is more common on window frames and sills, and it is found in a wide range of timbers. In New Zealand it is one of the major causes of decay in timbers used in buildings.

Surface damage	Softening of the timber, leaving it fibrous and stringy without substance. White rot in oak and other hardwoods
Identification	The fruiting body is upside down, with very thin layers of the mycelium below the pore zone. It is yellow-ochre in colour, sometimes darker or a dark brown colour. Microscopic identification is easy due to the presence of pointed brown form of hyphae
Rate of growth	Believed to be about 300mm per year
Dampness	22% – upper limit not recorded
Temperature	0–31°C

NON WOOD-ROTTING FUNGUS

* *Peziza* or 'Elf-cup'. The fruiting bodies may be seen in damp plaster or on saturated brick walls. They are fleshy pale yellow-orange, up to 50mm wide and break easily when handled. When they dry, they become hard and the colour deepens to a stronger reddish-orange. An indication of high moisture content.

* *Corprinus domesticus* or 'Ink-cap'. The mycelium has an appearance of coconut matting with slightly orange coloured tufts. It may appear as a yellow-orange sheet. The fruiting body is a slender toadstool up to 70mm high. The head of the toadstool is white and it eventually produces a black ink liquid in which the spores are freed. The fruiting body is short-lived and shrivels up onto the material which supported it. This fungus does not damage timber.

* *Myxomycetes* or 'Slime moulds'. Most moulds fall into one of four groups: *Penicillium*, *Trichoderma*, *Aspergillus* and *Pullularia*. They grow on damp surfaces such as plaster, brick, wallpaper, fabrics and clothing. They are commonly associated with areas of high condensation or in areas of moist humid air. Shower rooms and sports changing rooms frequently suffer from condensation, as do bakeries, breweries, and many factories manufacturing goods where a large water discharge occurs. The mould gives off spores very easily, and it is these spores that give the colour to the mould growth. These colours change from black to green with some greys, off white, and even yellow and pink. The *Aspergillus* group, which grow on damp patches usually on houses, can cause *aspergilloma* and *bronchopulmonary aspergillosis.* Young children and old people are both vulnerable to this infection. The slime moulds or *myxomycetes* are formed in delicate globular structures about 20mm in diameter. These are usually black in colour and will release black dust particles if broken. If you wipe them with the back of your hand they seem to leave a larger dirty mark.

Crack Interpretation

Nearly all buildings will have some cracking within them. The surveyor's role is to locate the failures that do exist and then to interpret the cause and the risk of future deterioration that may exist. The accessible evidence may be sufficient to confirm the cause or only suggest that a problem could exist. Cracking may be the result of moisture changes in the components of the building or the ground, deterioration in the construction or its content, chemicals within the construction, or the influences of climate.

Movement in buildings is an everyday occurrence. It may be the natural consequence of drying out, changes in temperature or humidity, varying movement in different materials, changes in load or the consequence of alterations that have been carried out. Buildings are now constructed to be stronger, but less flexible. The sight of fractures in a building is not always a cause for concern. A balanced appraisal will only be achieved by a greater awareness of the causes of cracking and the acceptance that not all cracks are caused by subsidence.

Table: Guide to cause of movement

Movement	Direction	Material	Notes
Post-construction movement due to moisture	Expansion	Brickwork	Plus 0.5mm/m within six months of construction completion
	Contraction	Concrete and calcium silicate	1mm/m within first two years of construction
		Timber	Movement within range of 20mm/m over first two years, speed influenced by climate and heating within building
Moisture-induced changes	Expansion	Water	Freezing of water increases volume by about 3%
Influenced by exposure of the building	Expansion	Metals	Rusting of steel products will increase volume by a multiple of up to four times in circumstances where the metal is under pressure
	Expansion	Chemicals in brick or concrete	Sulphates within construction materials expand when damp
	Expansion	Clay	Expansion due to introduction of water through pipe or drain failure, removal of trees or vegetation. Movement reversible
	Shrinkage	Clay	Shrinkage due to loss of water caused by climate or trees or vegetation. Moisture reversible.
	Shrinkage	Peat	Reduction following loss of moisture or addition of load. Movement can be substantial
	Shrinkage	Moving ground water	Water washed away ground leading to subsidence

Vegetation	Expansion	All	The growth of plants, shrubs and trees may influence the performance of a building. Roots growing into walls will expand joints (e.g. ivy in walls, buddleia roots in all materials)
Insect	Shrinkage	Timber	Root action beneath foundations can physically disrupt buildings; trees may strike parts of a building as they are moved in the wind. Contraction of soils is referred to above
Over-loading or load changes	Settlement	All	Beetle attack eating away at timber leading to structural failure or compression of support timbers
			Over-loading of a property will make it vulnerable to compression, or beams bending under the load. Such disruption will have a direct affect upon the building
	Progressive settlement	Mainly concrete	Some materials will continue to deflect even though the load remains constant
Ground failure	Settlement	Ground	Compression of fill material, compressible grounds
Mining, chemical extraction or quarrying	Settlement	Sub-ground	Collapse of layers within the ground due to the removal of part of the ground. Examples of this type of failure include: removal of stone (e.g. Cheshire); pumping chemicals out of the ground (e.g. Bristol); and extraction of coal (e.g. mining areas generally within the country)

The dynamic effect of temperature upon a building will be influenced by the coefficient of linear expansion for the component of the building and the expected temperature range that the material will experience. Temperature changes will depend upon the exposure of the material. Walls may experience a smaller temperature range than a roof. A dark material will soak up more heat than a light and reflective material. The table below sets out the coefficient for typical materials and the type of temperature range each may experience.

Table: Thermal movement range

Material	Cle/10⁻⁶	Temperature (°C)	Max movement (mm/m)	Lower range (mm/m)
Asphalt	30–80	85–105	8.4	2.55
Brickwork	6–12	70–85	1.02	0.42
Concrete	12–14	65–85	1.19	0.78
Glass	9–11	65	0.715	0.585
Glass coloured	9–11	115	1.265	1.035
Lead	30	85–105	3.15	2.55
Plaster	12–21	85	1.785	1.02
Polyethylene	160–200	70	14	11.2
Timber (across grain)	30–70	85	5.95	2.55

The performance of building materials following movement will depend upon their flexibility.

Table: Moisture and thermal movement range

Material	Reversible movement (mm)	Initial and non-reversible movement (mm)	Upper extent of ongoing movement (mm/m)	Extent of initial movement (mm/m)
Mortar	0.02–0.06	(−) 0.04–0.01	(+) 0.6	(−) 0.4
Concrete	0.02–0.06	(−) 0.03–0.04	(+) 0.6	(−) 0.4
Brickwork	0.02	(+) 0.02–0.010	(+) 0.2	(+) 0.2
Concrete blocks	0.03–0.06	(−) 0.02–0.10	(+) 0.6	(−) 0.6

RECORD TAKING

The investigation of the cause of cracking depends upon accurate records. You should record:

* The building's method of construction.
* The measurement of each crack width.
* The precise location of each measurement.
* The direction of movement; hogging, heave, subsidence, shear or horizontal movement, vertical displacement, rotation or horizontal displacement must be recorded.
* The presence of any displacement.

- The date of movement, or the probable period within which movement occurred.
- The condition of the crack.
- The materials that have cracked.
- The presence of any abnormal levels of moisture.
- The presence or absence of movement at junctions in materials.
- The presence or absence of cracking at the floor/skirting junction.
- The presence or absence of cracking going through the damp course.
- The nature of the ground.
- The presence of local watercourses, drains, rivers or trees.
- The horizontality and verticality of the construction roof and floors, and whether measured (if so, how) or estimated.

Annotated sketches of the building, and proportional plans, with the cracks shown in a contrasting colour, are a great help in deciding the cause of failure. Photographs can be a useful addition to the recording of information.

In new construction a variation in the level of walls of plus or minus 15mm may be part of the construction and not the result of post-construction movement.

Movement can identify the cause

In most cases one may expect the crack to be perpendicular to the line of stress.

THE EXAMINATION TO FIND CRACKS

Consider the implication of the following upon the risk of cracks being present.

- Quality of original construction – old buildings may have been built to a poor standard. The outer face may have been completed after the structure of the building and not bonded to the remainder of the building.
- The presence of a clay ground with foundations of less than 1m depth.
- Topography – a sloping site is more likely to result in movement occurring.
- Ground close to a river or a ditch will be more vulnerable.
- Cliffs may be damaged by the elements, or the combined effect of water seepage from drainage systems.
- The influence of global warming may affect the performance of a building.
- The presence of local mining, quarrying or extraction of chemicals.

- Proximity to major roads carrying heavy traffic.
- Water table levels and direction of water table level changes.
- The condition of pipes.
- The failure of water pipes, whether they are drains or service pipes, may cause the ground to move or be the consequence of movement that had already taken place. Tree roots may have entered the drain and their minor roots grown so that they fill the pipe bore.
- Proximity to major construction work involving deep foundations.

MONITORING

Monitoring a building is aimed at providing information, the interpretation of which should help in the accurate diagnosis of cause and effect. When movement occurs, and in what direction, should be investigated if the building sits on a clay site. The period of the monitoring is dictated by what is hoped to be learnt.

Seasonal influence on a clay ground can be established within a few months of starting to monitor. If you start in later spring it may take only four months to confirm that the pattern of crack movement matches that expected where a building is located on shallow foundations in a clay ground.

Short-term monitoring of a crack, for example, between September and November, will show if there is any reduction in crack dimension. That movement could be due to:

- a pipe leak wetting the ground and causing expansion;
- drainage failures below the ground causing expansion;
- changes in the water table affecting the volume of the ground;
- changes in water content of the ground, caused by diversion of water courses or flooding;
- damp in timbers within the building causing them to expand;
- timber expansion prior to the start of the central heating season; or
- seasonal changes in water content resulting in heave in the ground.

The checking of cracks between May and September may show that they become larger for the reverse of the reasons outlined above.

To monitor cracks, measure each crack on a regular basis. The monitoring must enable the amount and direction of movement that has occurred within the period between measurements to be identified. I suggest that the building is photographed each time that the monitoring is recorded.

Crack measurement should record:

* the crack width;
* any displacement; and
* any lateral movement.

Measurements should be made onto a blank form, and transferred onto the continuous record after measurement has been completed. This is to avoid auto-suggestion of the answer. Where readings appear to be in conflict, readings should be checked.

Monitoring a crack is done for one of three reasons:

* to see which part of the building moves and when;
* to see by how much the building moves and recovers; or
* to see what is the rate of movement.

The use of tell-tales may tell you how much movement has taken place, but they record the movement in only two dimensions. They will not record which side of the crack moved, or whether any displacement occurred, and do not give a sufficiently precise measurement for record purposes.

RESEARCH

Ground investigation

In order to decide what might have caused the movement in a building, a thorough investigation is required. This may include:

* The determination of the construction and materials used.
* The determination of the quality and state of the materials of the building.
* The determination of the anticipated thermal and moisture related movements for each part of the building.
* Digging trial holes to inspect the foundation depth and the quality of the ground beneath.
* Establishing the water content of the ground.
* Bore or auger tests.
* Drains test.
* Water service pipe test.
* Petrographic analysis of cracks – examination of a thin slice through the material that is cracked.
* Geological maps – but note the limitation in their accuracy when using them to identify the ground below a single building.
* Old maps – to determine what may have been on the site before.
* Aerial photographs.

Table: Crack diagnosis

Crack shape	Possible cause
Horizontal Repeated at regular intervals in brickwork or render on brickwork	• Corrosion of wall ties or other metals within the wall. This is usually more extensive on the exposed sides of the building. There should be an equal number of brick courses between each horizontal crack. If rendered, the repeat cracks should be an equal distance apart • Check by examining carefully the wall where the next failure is predicted
Horizontal Single horizontal joint crack	**High level** • Roof spread. The roof rotates the outer brickwork as the load is taken on the outer leaf of a cavity wall. This causes a crack two or three brick courses below the eaves. Look at the roof covering for recent use of concrete tiles • Wall tie corrosion because it increases the height of the outer leaf of the wall • Deterioration of the purlins within the roof or the absence of diagonal bracing • High wind vibration can also cause roof movement if there are no holding down straps
Horizontal Single horizontal joint crack	**Above windows** • Concrete lintels, where there is a difference in the coefficient of expansion between the concrete and the brick • Deterioration of brick slips set onto concrete lintels over the windows • Movement on the slip plane afforded by the cavity tray due to thermal influences. This will be allied to the presence of some vertical fractures • Timber bressummer failures behind a brick arch. May show as a horizontal crack where the lintel bends as it decays. This leaves a horizontal crack on the line of the top of the failed lintel

Continued

- The collapse of a soldier arch above a window can leave a horizontal fracture immediately above the beam
- Where brickwork has been restrained at its ends, may resist expansion. This may cause the restrained wall to arch upward leaving a long horizontal split. This is common with parapets, particularly if they have been rebuilt on top of a wall constructed with a lime mortar
- The presence of corroding steel lintels over windows

Horizontal
Single horizontal joint crack

Parapets
- Thermal movement. Common on a north elevation where the parapet wall is heated from the rear, which faces south. This enlarges and a slip crack occurs between this and the cooler elevation below. If the parapet is not straight, but steps in and out, vertical cracks will be found at the change of direction

Horizontal

Damp proof course
This forms a slip plane. Movement may be caused by:
- Chemical reaction (sulphates)
- Thermal movement
- Moisture expansion after construction
- Subsidence

Vertical
Regular spaced at 7–10m

The vertical crack is often related to thermal expansion or initial drying shrinkage. It will be related to the inadequate provision of movement joints. The location may affected by:
- Weakness within the wall envelope. e.g. insubstantial panels below windows
- Displacement, placing pressure on the wall close to a corner
- Different materials used in the construction having varied coefficients of thermal expansion
- Brick expansion after construction

Table: *Continued*

Crack shape	Possible cause
Vertical At wall junctions	Uneven movement between parts of a building supported in different ways. These cracks should be larger at the top than at the bottom if rotation is taking place. For example: • At the junction with a rear extension or outrigger • Where the main house joins onto a wing of the building. If the wing subsides, the junction becomes a hinge and a vertical split takes place at this junction A vertical crack is inevitable if there are different materials forming the walls that join. Brickwork and blockwork will shrink by different amounts Reinforcement corrosion in concrete lintels or beams, copings or other reinforced concrete may give a straight fracture in the material. It will be horizontal or vertical, depending upon the direction of the reinforcement
Vertical At junction with a bay window	• Water seepage may have resulted in structural timbers of the bay deteriorating, resulting in any cladding pulling away from the adjoining walls and the timber frame. It can look like subsidence in the bay but it is settlement of the components of the bay • Upward pressure from foundations. Because the house is heavier than the bay, the house will push down on the foundations. The result is that the house subsides to a point where the ground has compressed sufficiently to carry the load but the bay is pushed upward. Cracks at the junction of front wall and bay may look like a tear and may suggest the bay has moved as opposed to the house • The bay may have foundations at a higher level than those to the house. They may be vulnerable to the shrinkage of the subsoil at that level. In such a case the bay rotates away from the house, leaving a vertical crack, but of varying thickness

Vertical
To the side of a window

- Poor window fixing
- Window loose
- Different coefficient of expansion in materials of wall and window, e.g. timber, plastic coated metal, metal painted
- Subsidence, because this is a weak part of the wall

Vertical
Within walls and to corners

- Independent addition of front brick skin in Victorian and Georgian buildings
- Use of rat trap bond, or snapped headers
- The use of mathematical tiles

Diagonal
At corners

- Commonly linked with movement in the ground. This could be caused by erosion, expansion, changes in the water table or similar occurrences dealt with above. Whilst this is the most obvious cause, it is not the only one
- Caused by the stress at right angles to the crack. This may be caused by the corner being pulled down due to foundation failure or the corner being pushed upwards, because of heave
- The type of mortar will affect the vulnerability of the wall to movement
- In a frame building the reciprocal of this lack of support can be vertical and horizontal cracks at joints and alongside infill panels
- A raft foundation may act as a single unit so that the cracking occurs at junctions with attached buildings
- The movement may have occurred to the adjoining building, e.g. an attached garage that has been inadequately founded moves, and acts as a lever at its connection with the second building

Continued

Table: *Continued*

Crack shape	Possible cause
Diagonal At corner	• Movement in the ground • Slippage of the ground on slope • Erosion of ground, by flooding, water movement in ground or failure of water pipe or drain • Compaction due to vibration • Partial overloading
Diagonal	• Impact damage on the corner by vehicle. The movement tends to end on a slip plane, often being the horizontal damp course
Diagonal Full height	• **Rotation on a sloping site.** An earthquake may vibrate the soil so that it liquefies and flows down the slope. Weight on a slope, a building or a terrace may contribute to slide • **Subsidence of magnitude** due to compression of de-watered peat or clay shrinkage. Clay may shrink and affect foundations to a depth of 3m, but foundations less than 1.5m deep are most at risk
Diagonal or vertical Part height	• **Lack of restraint in a flank wall** – e.g. where the staircase runs alongside a flank wall or floor joists run parallel and are not tied to the wall. Look for a bulge in the wall • **Roof spread** – allied to a bulge on the wall and displacement in the wall materials
Crazing In render	May be the result of: • the application of the render in too thick a coat • too rapid drying out of the render • chemical additives, including sulphates in render • application of render to fletton bricks

Cracks both sides of wall

Cracks on both the inside and outside of a cavity wall

Movement has affected the wall and not just the outer face or the inner face. Thermal or drying shrinkage can be ruled out. The most likely cause of the movement will be foundation failure or the ground having dropped

Cavity wall typical causes of cracking

- Expansion due to moisture wetting/thermal movement
- Sulphate attack to walls and floors from soluble salts expanding
- Corrosion or oxidation of steel (wall tie repairs)
- Moisture expansion of fire clay products which is caused by inappropriate location of movement joints
- Carbonation causing shrinkage cracking in some blocks and volume changes causing internal shrinkage cracks
- Foundation movement impaired load changes within the structure due to settlement or movement

More than one cause

Close examination will often show that the cracking is due to more than one failure. It may be, in part, due to the normal shrinkage, partly due to thermal movement, and the remainder due to a small amount of subsidence. For this reason the surveyor must allow the possibility of there being more than one form of movement.

THE TYPE OF GROUND

1. **Rock.** Should require a pick to make an impression upon the ground. Foundations to internal and external walls should be at least 300mm deep.
2. **Gravel or sand.** Gravel will have a granule size between 2–4mm. Sand will have granules of between 1–2mm. The foundation should be at least 600mm wide and to a depth of at least 300mm for internal and external walls.
3. **Clay, sandy.** Clay is the most common ground materials to influence the movement in buildings, can expand and contract by as much as 10% by volume. Most winter rainfall will soak deep into the ground, whereas summer rainfall runs off the surface or evaporates. Sand will have granules of between 1–2mm. Clay tends not to have grit when tested on the tongue. Foundations are 600mm wide and a minimum of 1.2m deep, but 3m if close to trees.
4. **Silt or clay.** Granules of between 0.002mm–0.06 – for silt, clay tends not to have grit when tested on the tongue. Special foundation design required.
5. **Peat or made-up ground.** Cracks are more likely to be present in buildings that sit on weak or vulnerable ground. The peat of the Fenland of Lincolnshire, Cambridge, Suffolk and Essex is very compressible. Recent action by local River and Water Authorities to extract water from rivers has reduced the level of the water table in many Fenland locations. Little or no expansion risk. Special foundation design will be required in these circumstances.

SETTLEMENT AND SUBSIDENCE

Meaning

+ 'Settlement' is applied to the failure of the components of the building; and

• 'Subsidence' is used where the building is damaged by a failure in the ground.

Insurance cover for subsidence

• The cost of repair (or loss in value).
• Where the damage was caused whilst this insurer was 'on risk'.
• Pre-existing damage started before this insurer took on the risk is not covered.

Advice about cracking could include recommendations for:

• Further investigation to be carried out before commitment to purchase.
• A warning of the risk of further movement.
• Guidance as to the anticipated problem.
• The purchaser takes an assignment of the existing insurance policy.
• The purchaser takes an assignment of the benefit of a claim to be made by the vendor.
• The extent of any excess under the insurance.
• Any risk of under insurance.
• The damage caused by building operations.
• The financial risk of being responsible for building works.
• The possibility of the works overrunning the contractors time and financial estimates.

In advising on a building's faults and their implications, the alternatives from which the surveyor must select an option are:

• Not serious and can be ignored.
• May become serious in the future.
• Serious at the moment.
• Urgent – repair required now.
• Uncertain as to the risk, therefore further investigation needed, especially before any commitment to purchase.

INFLUENCE OF TREES

In ground that is vulnerable to shrinkage through moisture change:

• Downward movement where there is a new tree.
• Upward movement where trees were removed when building constructed.

* If a structure settles evenly little damage is caused.
* No damage should occur where the ground is dry near to trees because water content should not change at foundation level unless a tree is removed.

Tree roots can vary the moisture content of the ground for depths of over 4m. There is little risk where the tree is its mature height from a building.

Damage to property or disturbance to occupants may have resulted from a number of causes:

* root action causing ground shrinkage;
* root action causing pressure in retaining walls or foundations;
* root action causing failure in drains;
* damage from falling leaves blocking gutters, drains or water gullies;
* tree branches striking roof surfaces or parts of the building causing damage;
* falling trees;
* trees seeping sticky substances which stain paths, cars, driveways, etc.
* expenditure incurred in clearing up leaves each year;
* creeper action causing mortar failure in brickwork and masonry;
* creeper action causing failure to the face of weak block or brickwork; and
* creepers blocking outlets and openings such as gas ventilators and air bricks.

Where the building is near existing trees, the type of each tree should be identified so that one can consider the influence of this tree in the specific ground conditions. Plot the trees' position on a scale plan of the site. Show the probable radius of influence by drawing circles based upon the mature height and existing tree height. The influence of the trees can then be estimated both now and in the future. Re-inspect the vulnerable parts of the building.

The roots of trees can spread over considerable distance. General indications suggest that trees are unlikely to cause damage where they are the distances set out in the table below from a vulnerable building.

Table: Recommended distance of trees from a building by type

Up to 10m	10–15m	15–25m
Acacia	Horse Chestnut	Elm
Apple	Lime	Oak
Ash	Maple	Poplar
Beech	Plane	
Birch	Sweet Chestnut	
Birch	Sycamore	
Cherry	Willow	
Cypress		
Damson		
Hawthorne		
Holly		
Hornbeam		
Laburnum		
Larch		
Laurel		
Magnolia		
Pear		
Pine		
Plum		
Rowan		
Spruce		
Walnut		
Yew		

Single trees are less likely to be the cause of clay movement below shallow foundations at the above distances. Where there is more than one tree the risk increases.

The location of the property will affect the risk of ground movement in clay soils:

• Built in open ground away from major vegetation – low.
• Old building near existing trees – low/moderate.
• Built on a site where trees have recently been removed – high.
• New building on site of mature trees – low.
• New building with recent planting – high.

Floor slab failure may be due to:

• Inappropriate or poorly compacted under floor fill.
• Chemical attack by sulphates in hardcore.
• On sloping sites, instability of the ground below a slab.
• Movement of water through the ground.

The risk of continuing movement

Where made-up ground occurs, or there are defects in the ground near the building, the failures usually occur within the first ten years of a property's life.

REPAIR

Underpinning has been considered to be a remedy where movement has taken place. It has not been a universal success. Much mass fill foundation repair was inadequate, either in design or construction. The surveyor must advise upon the implication of the repairs, not just the cost, but also the time scales and the disruption that may be involved.

The most common methods of repair are set out below.

Foundation repair options

* Absence of consent – foundation repair requires consent under the building regulations. Where the work may influence an adjoining building consent may not be forthcoming. E.g. where repairs are proposed to the foundations of a building that is part of a terrace.
* Guarantees are usually of little value – they may be limited to the cost of the foundation repair, with no provision for the upgrade of costs for inflation in the future. They may not cover the cost of superstructure repairs or the provision of alternative accommodation if remedial work is required. It is common for the guarantee to avoid liability for heave. They rely upon documents to be enforced. Often depend upon the company remaining solvent.
* Mass fill underpinning – 1m-wide bays are excavated, the ground is removed and concrete poured in to fill the void up to just below the existing foundations. This is filled with a dry fine concrete. Excavations must not undermine more than 20% of an elevation at a time. New bays are not placed alongside recent work until five days have passed. Failures are common. May adversely affect the value of the building.
* Partial underpinning – in special circumstances a small part of the building, e.g. a corner, may have the foundations strengthened. There remains the risk that this plug could settle as the ground takes up the new load. May adversely affect the value of the building.

- Mini piles – this repair involves the installation piles with diameters between 65–150mm. They are either driven through the existing foundation at regular intervals or linked to the original foundations by cantilevered pile caps. The type of ground will determine the design of the pile method. This system of foundation repair is suitable for buildings that have suffered from ground compression, such as would occur where the ground contained vegetable matter. The system is not suitable where the building has been subjected to substantial amounts of subsidence, or where there is a risk of heave. This system is only suitable if the foundations of the building are in good condition.

- Injection – the injection under pressure of a cement grout into the ground. The weakness of the system is the difficulty of making sure that the concrete grout is pumped into the void you require filled. This method of repair may be appropriate for the replacement of ground support to a car park, or even a ground slab.

- Stiffening beam – a beam placed in the wall above the foundation, being cast in 1m-wide bays. Avoids the cost of excavating below the foundations and minimizes the extra loading on the weak ground. May form a ring beam that can be linked to pile foundations at a later date, as a second stage to a repair. Not suitable if there is a risk of heave, or substantial ground movement.

- Pier and beam – excavation of piers and casting a reinforced concrete ring beam above or below the original foundations of the property. Piers maximum depth about 4m. Cheaper system of underpinning than mass concrete, if the depth of the foundations is below 2m. Causes less damage to the superstructure. Suitable in clay soils. If internal walls require support it becomes disruptive to install.

- Pile and beam – instead of digging out the piers, piles are driven into the ground, to depths below 4m. The piles are usually 200–400mm in diameter and can go to depths in excess of 15m. The disadvantage of driven piling is the vibration created in an already vulnerable building. The vibration may also damage adjoining property.

Consider the guidance that is given by the BRE in Digest 251 about interpreting the damage to a building.

Table: A guide to possible damage that has been caused based on static crack dimension

Category of damage	Degree of damage	Crack width (mm)	Description of the crack
0	Negligible	Up to 0.1	Hairline cracks of less than 0.1mm width are difficult to see in the external walls of a building. In brickwork they can easily be overlooked. Decide if there is a risk that the crack may be progressive
1	Very slight	Up to 1	Fine cracks which can be treated easily during normal decoration or redecoration. The damage is very slight and would have no effect on the building's use
2	Slight	Up to 5	Cracks can easily be filled and redecoration will probably be required. Recurrent cracks may have to be masked with a suitable lining. Pruning or pollarding offending shrubs or trees may deal with seasonally reoccurring cracks. Pruning does encourage root action. Monitoring should be put in hand to identify the rate of progression of the cracks present. Underpinning is unlikely to be cost-effective for this level of cracking

3	Moderate	5–15	The cracks can be patched by a bricklayer or stonemason. Doors and windows stick and there is a risk of service pipes fracturing. Weather tightness is often impaired. Vulnerable brick arches or lintel bearings need close examination and propping may be needed. Monitoring should be used to establish the rate of progress of the movement. Underpinning may or may not be cost-effective for this level of cracking
4	Severe	15–25	This will require extensive repair work involving breaking and replacing sections of walls, especially around doors and windows. Window and door panels are distorted, floors slope, walls lean or bulge. The bearing of beams may be affected and service pipes are disrupted. Stability is now at risk. Monitoring should be used to establish the rate of movement. Unless the cause of the movement can be easily remedied underpinning will be required
5	Very Severe	Over 25mm	Major repair required involving partial or complete rebuilding. Beams lose bearings, walls lean badly and require shoring, windows broken with distortion. There is a danger of instability. The inspection of the building will require care, as part of it may be unstable. Steps need to be taken quickly to prevent further damage in the event that progressive movement takes place

WARNINGS

* Do not rely on crack dimension alone in considering the risks.
* Remember that the crack that caused a building to collapse started small.
* Crack dimension must relate to the frequency of cracking. For the sake of this interpretation allow the dimension of a crack to be the combined width of all cracks within, say, 2m.
* Displacement between one face of a wall and the other, to each side of the crack must be added to the crack width.

BULGES

Vertical bulging may be remediable or allowed to remain where stable when the distortion does not exceed one sixth of the thickness of the wall. Where the distortion exceeds that amount, the wall will require careful examination and calculations to determine whether it can be saved. When the movement exceeds a third of the wall thickness it is almost certain that the wall will have to be rebuilt or additional buttressing added.

RISK EVALUATION

Reach your decision on the basis of what can be seen, on the consequences of the crack and the cause of the cracking. There will be several options available when it comes to reporting to the client.

* The crack is either static or dynamic.
* The crack is either a representation of the full extent of the movement that will take place, or the extent of the movement that has taken place, but is no guide to the extent of the future movement that will occur.

Focus

WHAT IS A SURVEY? WHAT ARE YOU DOING?

The inspection is observation, the report considered opinion.

1. **Look at the condition of what is there.**
   - What level of inspection and test have you agreed?
   - Tests prove the condition or lack of defect at the time of the inspection. Notes confirm the data collected at the time of inspection.
   - What can you prove?
   - Advise on the basis of probability.
2. **Look for what should be there.**
   - Advise upon the consequence of omission. Ensure that you have an adequate knowledge base of the type of property being inspected. Know limitations or extra duty for leasehold or commercial property.
   - Inform on liability.
3. **Consider the condition of what you cannot see.**
   - Advise upon the risk that may be present, and inform how that risk can be rationalized; for instance, by testing services, opening up suspect and concealed areas, before exchange of contract.
   - State assumptions made.
4. **Record tests and checks.**
   - Add the results of your checks, for instance the size of any cracks, the level of damp recorded, the extent of floor slopes, etc.
5. **Inspection.**
   - What was the agreed level of inspection:
     assumed
     contracted
     imposed
     at the date of the survey?
   - What is your proof of the information available to you at the time of the inspection?
   - What equipment was used, and what could it reasonably have discovered?

6. **Warning.**

Knowledge at the time of the inspection:

- published by the Building Research Establishment;
- published by the professional bodies;
- established as common knowledge by local surveyors;
- based upon identification of type of construction; and
- logic, the projection of circumstances.

Report:

- On the consequences of defects present, or believed to be present.
- On the ease or complication of repair. Give an indication of range of cost.
- On the probability of major expenditure after the first year of occupation.
- What the client should do before they commit themselves to the purchase.
- If the client buys, what they have to do before they move in.
- The report should be in a format to be read by the common man before the exchange of contracts.

Index